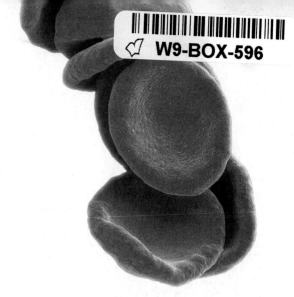

Human Physiology

FIFTEENTH EDITION

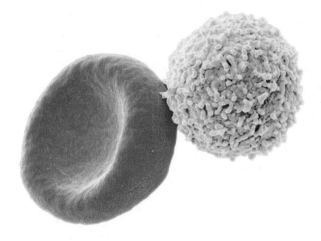

STUART IRA FOX
Pierce College

Krista Rompolski
Drexel University

McGraw Hill Education

HUMAN PHYSIOLOGY

Published by McGraw-Hill Education, 2 Penn Plaza, New York, NY 10121. Copyright © 2019 by McGraw-Hill Education. All rights reserved. Printed in the United States of America. No part of this publication may be reproduced or distributed in any form or by any means, or stored in a database or retrieval system, without the prior written consent of McGraw-Hill Education, including, but not limited to, in any network or other electronic storage or transmission, or broadcast for distance learning.

Some ancillaries, including electronic and print components, may not be available to customers outside the United States.

This book is printed on acid-free paper.

2 3 4 5 6 7 8 9 LWI 21 20 19 18

ISBN 978-1-260-09284-4
MHID 1-260-09284-4

Cover Image: ©*Science Photo Library/Alamy Stock Photo*

The Internet addresses listed in the text were accurate at the time of publication. The inclusion of a website does not indicate an endorsement by the authors or McGraw-Hill Education, and McGraw-Hill Education does not guarantee the accuracy of the information presented at these sites.

mheducation.com/highered

Brief Contents

Stuart Ira Fox earned a Ph.D. in human physiology from the Department of Physiology, School of Medicine, at the University of Southern California, after earning degrees at the University of California at Los Angeles (UCLA); California State University, Los Angeles; and UC Santa Barbara. He has spent most of his professional life teaching at Los Angeles City College; California State University, Northridge; and Pierce College, where he has won numerous teaching awards, including several Golden Apples. Stuart has authored forty editions of seven textbooks, which are used worldwide and have been translated into several languages, and two novels. When not engaged in professional activities, he likes to hike, fly fish, and cross-country ski in the Eastern Sierra Nevada Mountains.

To my wife, Ellen; and to Laura, Jacob, and Kayleigh. For all the important reasons.

© Ellen Fox

Krista Lee Rompolski earned her Ph.D. in exercise physiology from the University of Pittsburgh, Department of Health and Physical Activity, after earning her bachelor's and master's degrees from Bloomsburg University, near her birthplace of Mount Carmel, PA. Since completing her Ph.D., Krista has been teaching Anatomy and Physiology, Pathophysiology, Exercise Physiology, and clinical research courses at Drexel University in Philadelphia. Krista is actively involved in continued anatomy and physiology education, academic research, and physical activity research. When not engaged in professional activities, she enjoys dancing, reading, and traveling anywhere with a rich history.

To my parents, Elaine and John, who taught me how to chase my dreams and believed in me first; and to my husband, Dan, my very best friend, who picked up right where they left off.

© Katherine Coccagna

Preface

The Story of the Fifteenth Edition

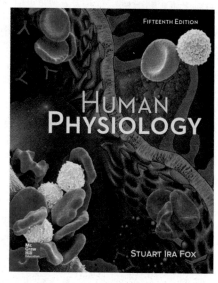

Stuart Fox, Ph.D., wrote the first edition (published 1983) to help students understand the concepts of human physiology, and this objective has remained the guiding principle through all of the subsequent editions. All editions have been lauded for their readability, the currency of the information, and the clarity of the presentation. The fifteenth edition continues this tradition by presenting human physiology in the most current, readable, and student-oriented way possible. This milestone edition is marked by a unique cover, the addition of a Digital Author, a new art program, and the updating of terminology and content.

It takes a village! To create this landmark fifteenth edition, Stuart had the support of Krista Rompolski as the Digital Author and a superb team at McGraw-Hill Education and MPS Limited. This team includes Michael Ivanov, Fran Simon, Andrea Eboh, Kelly Hart, Jessica Portz, Christina Nelson, Joan Weber, Angela FitzPatrick, Amy Reed, Jim Connely, Kristine Rellihan, Matt Backhaus, and Lori Hancock. We are all incredibly grateful to the many reviewers who provided their time and expertise to critically examine individual chapters and be Board of Advisor partners. These reviewers and advisors are listed on the pages that follow.

Cover

The cover is the most immediately apparent change for the fifteenth edition. For this edition, Bill Westwood's cover art depicting the glomerular capillaries of the kidneys from the 1983 first edition of the book serves as the background, while the foreground features colorized scanning electron micrographs of blood cells. Westwood, a renowned biomedical illustrator, created all but one of the covers of the previous editions.

Digital Author

For the first time, this textbook has a Digital Author, Krista Rompolski, Ph.D. (Drexel University). Krista Rompolski is responsible for all of the digital content that supports the fifteenth edition, ensuring that it is specifically tailored to the needs of students and instructors who use the textbook.

As part of this effort, Krista collaborated closely with Stuart, carefully reading every line of the text and making numerous suggestions and comments, many of which were adopted. This partnership greatly aided the author in ensuring that the fifteenth edition became even more current, readable, and student friendly. As a result of Krista's significant investment, the digital resources for the fifteenth edition are seamlessly integrated with the textbook content and will continue to be a uniquely useful tool for students and professors.

New Art Program

Every piece of art has been re-rendered for this landmark fifteenth edition! The goal was to make every figure more accurate, clear, vibrant, and readable. By comparing the fifteenth edition to previous editions, the eye-popping beauty of the art program is immediately apparent. Smaller changes, such as an updating of many anatomical terms, may not be as readily apparent, but will serve to better aid student learning.

The stunning, instructional figures will increase students' enjoyment as they use this textbook, and will increase their motivation to learn the concepts covered in the course. An added pedagogical element is the "location" icon that appears on many figures. This icon serves to help students better integrate cellular and molecular processes with their locations in body organs.

Terminology and Content Updates

Stuart and Krista would like to extend special appreciation to Beth Kersten, State College of Florida, for her careful evaluation and contributions to making terminology update recommendations. Terminology was updated throughout to be consistent with what is used by industry professionals and to also be compliant with *Terminologia Anatomica, Terminologia Histologica, Terminologia Embryologica, IUPAC Nomenclature of Organic Chemistry,* and *IUPAC Principles of Chemical Nomenclature.*

As with all editions, the content updates made by Stuart are in line with the latest research and are presented with the breadth and depth that is appropriate for the undergraduate student of human physiology. In addition to the entirely new art program and updated terminology, approximately 31 new and 92 updated discussions have been integrated. A detailed list of these changes is listed on the pages that follow.

Guided Tour

WHAT MAKES THIS TEXT A MARKET LEADER?

Clinical Applications—No Other Human Physiology Text Has More!

The framework of this textbook is based on integrating clinically germane information with knowledge of the body's physiological processes. Examples of this abound throughout the book.

CLINICAL INVESTIGATION

Sheryl, an active 78-year-old, suddenly became greatly fatigued and disoriented while skiing. When she was brought to the hospital, blood tests revealed elevated levels of LDH, AST, ALT, and the MB isoform of CK.

Some of the new terms and concepts you will encounter include:

- Enzymes, isoenzymes, coenzymes, and cofactors
- LDH, AST, ALT, and CK

▶ **Clinical Application Boxes** are in-depth boxed essays that explore relevant topics of clinical interest and are placed at key points in the chapter to support the surrounding material. Subjects covered include pathologies, current research, pharmacology, and a variety of clinical diseases.

EXERCISE APPLICATION

Metabolic syndrome is a combination of abnormal measurements—including central obesity (excess abdominal fat), hypertension (high blood pressure), insulin resistance (prediabetes), type 2 diabetes mellitus, high plasma triglycerides, and high LDL cholesterol—that greatly increase the risk of coronary heart disease, stroke, diabetes mellitus, and other conditions. The incidence of metabolic syndrome has increased alarmingly in recent years because of the increase in obesity. Eating excessive calories, particularly in the form of sugars (including high fructose corn syrup), stimulates insulin secretion. Insulin then promotes the uptake of blood glucose into adipose cells, where (through lipogenesis) it is converted into stored triglycerides (see figs. 5.12 and 5.13). Conversely, the lowering of insulin secretion, by diets that prevent the plasma glucose from rising sharply, promotes lipolysis (the breakdown of fat) and weight loss.

▼ **Learning Outcomes** are numbered for easy referencing in digital material!

LEARNING OUTCOMES

After studying this section, you should be able to:

2. Describe the aerobic cell respiration of glucose through the citric acid cycle.
3. Describe the electron transport system and oxidative phosphorylation, explaining the role of oxygen in this process.

CLINICAL INVESTIGATIONS IN ALL CHAPTERS!

◀ **Chapter-Opening Clinical Investigations, Clues, and Summaries** are diagnostic case studies found in each chapter. Clues are given throughout and the case is finally resolved at the end of the chapter.

CLINICAL APPLICATION

When diseases damage tissues, some cells die and release their enzymes into the blood. The activity of these enzymes, reflecting their concentrations in the blood plasma, can be measured in a test tube by adding their specific substrates. Because an increase in certain enzymes in the blood can indicate damage to specific organs, such tests may aid the diagnosis of diseases. For example, an increase in a man's blood levels of acid phosphatase may result from disease of the prostate (table 4.1).

◀ **Exercise Application Boxes** are readings that explore physiological principles as applied to well-being, sports medicine, exercise physiology, and aging. They are also placed at relevant points in the text to highlight concepts just covered in the chapter.

▼ Learning Outcome numbers are tied directly to **Checkpoint numbers!**

CHECKPOINTS

2a. Compare the fate of pyruvate in aerobic and anaerobic cell respiration.

2b. Draw a simplified citric acid cycle and indicate the high-energy products.

3a. Explain how NADH and $FADH_2$ contribute to oxidative phosphorylation.

3b. Explain how ATP is produced in oxidative phosphorylation.

DIGITAL UPDATES

Krista's Story:

In an age when many online learning tools are not only available but often students' preferred method of learning, quality of content is of the utmost importance. Evidence-driven texts, written by experienced physiologists should provide the foundation for such content. However, digital resources can confuse or waste the valuable time of students and instructors if they are not reflective of the textbook assigned for the course.

As the first digital author of Fox *Human Physiology,* my task is not only to supplement the textbook with a variety of learning resources and assessment methods, but to ensure that the digital content reinforces and highlights the strength of this text. This could only be accomplished by being thoroughly absorbed in the text. I spent the majority of 2016 participating in the updates to the text for the fifteenth edition with Stuart Fox. I learned Stuart's unique way of presenting the material in each chapter, telling a story, and reinforcing real-world application. I understand his vision, the tone of the text, and the intended audience. I also know where students may struggle, may need additional resources, or may need to spend more time on self-assessment to ensure mastery.

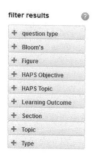

As a professor, I know how important it is to deliver quality assessments to students, and for those assessments to challenge them on a variety of levels. Thus, evaluating the assessment questions available to both instructors and students was a priority. To best meet the needs of instructors, we collected vast amounts of data on the historical usage of the available questions in Connect. This allowed me to focus in on revising rarely used questions or editing them to provide a greater variety of difficulty levels. Thus, regularly utilized questions remain available, and underutilized questions were revised. Even when questions are well written, digital question banks often lack variability in difficulty level and do not match the terminology of the text. Moreover, assessments in collegiate physiology courses should be based on the standards for anatomy and physiology education. To address these concerns, each question in Connect has been revised to match the terminology in *Human Physiology,* fifteenth edition, tagged by a learning outcome from *Human Physiology,* as well as from the Human Anatomy and Physiology Society (HAPS), and assigned the appropriate level of difficulty on Bloom's taxonomy. Furthermore, several new questions at level 3 or more on Bloom's Taxonomy have been added, for an even distribution of difficulty level across the question banks. These enhancements will allow instructors to create the most effective assessments possible.

Quality assessments only matter when students are well-prepared to tackle them. To accomplish this task, new case studies, videos, and active learning exercises have been added to

Connect for each chapter of *Human Physiology,* fifteenth edition. The Clinical and Exercise Applications from the textbook will be available in Connect, as well as new Connect-exclusive ones for students who want to investigate a topic further and apply their learned knowledge to real-world scenarios. The SmartBook application, our digital text, can provide instructors with feedback on which topics students are struggling to understand, and allow them to emphasize and clarify those topics in class. It can also assess students' reading comprehension in real time, with newly updated questions that appear as they read.

Digital resources that complement the text help to create consistency across the text, classroom activities, and the digital content. I have examined every question, exercise, and resource in Connect, with the textbook open by my side, to ensure a seamless flow between the two. I believe in this textbook, and am honored to help continue its legacy of success into the future. To ensure ease of adoption and usage, I will be available to instructors to answer questions and provide advice about the digital resources. I hope you enjoy *Human Physiology,* fifteenth edition!

- Expanded learning resources to help students understand more traditionally challenging concepts
- New test questions in each chapter, touching on all learning objectives
- Questions available on all levels of Bloom's taxonomy for each chapter, with at least 30% of the questions at level 3 or higher
- New and improved probes in SmartBook that allow students to assess comprehension as they read, and that provide feedback to instructors on areas of students' mastery or difficulty
- HAPS learning objective available for every question to ensure consistency with A&P standards
- Explanations for answers provided to students for all Bloom's levels 3 and above
- All questions in both test and question banks are tagged by topic, learning outcome, and HAPS learning outcome
- All test and question bank terminology and tone derives directly from *Human Physiology,* fifteenth edition
- New learning resources in a variety of formats, such as videos, interactive case studies, and labeling exercises, have been added to reinforce topics students traditionally find challenging

Dynamic New Art Program

Every piece of art has been updated to make it more vibrant, three-dimensional, and instructional. The authors examined every figure to ensure it was engaging and accurate. The fifteenth edition's art program will help with understanding the key concepts of anatomy and physiology.

Vibrant, bright, dynamic colors!

Location icons on relevant figures help orient students to understand where a physiological process occurs in the body as a whole.

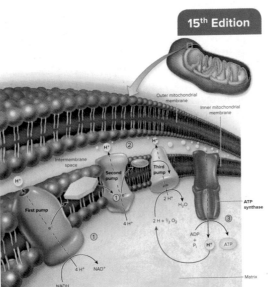

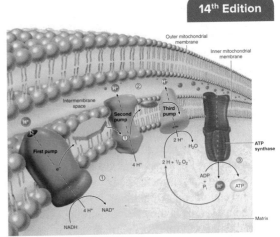

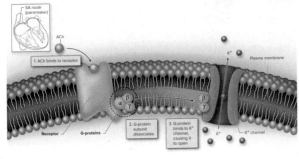

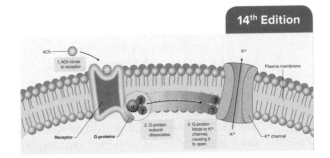

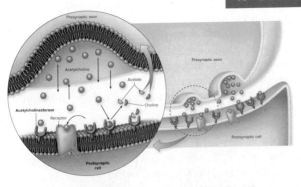

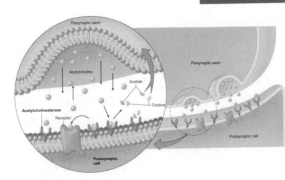

Many unique figures grace the pages of Fox: *Human Physiology* which aren't found in competitor titles.

Macro-to-micro images give students a closeup view.

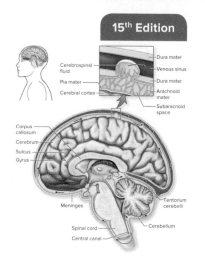

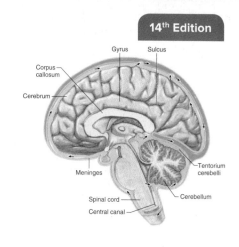

Reimagined illustrations are more relatable to clinically-focused students.

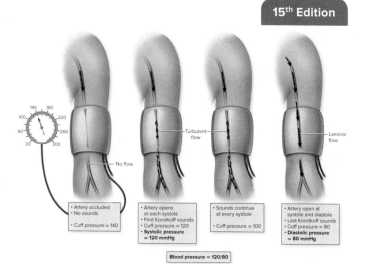

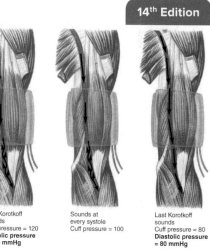

An inset illustration of a lung provides the connection that carbon dioxide is transported by the blood to the lungs, and processed for elimination.

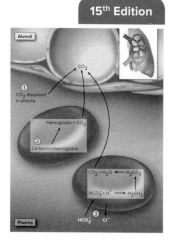

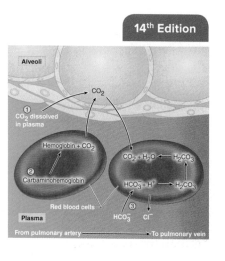

New to This Edition

UPDATED TEXT INFORMATION IN THE FIFTEENTH EDITION

There have been many changes in all of the chapters that are not included on this list. Numerous smaller changes, updated terminology, and all changes in art are excluded from this list. Some of the more important changes in concepts, facts, and explanations in the fifteenth edition include the following:

Chapter 1
1. Updated table of physiological discoveries (table 1.1).
2. Updated normal range for blood glucose (table 1.2).

Chapter 2
1. Teratogenic effects introduced along with medical uses of thalidomide.
2. Definition of carbohydrates and general formula for monosaccharides rewritten.

Chapter 3
1. List is provided of the different functions of plasma proteins.
2. Updated and expanded discussion of the primary cilium.
3. New discussion of CRISPR-Cas9 added.
4. New discussion of mutations and their roles in cancer added.
5. Section on epigenetic inheritance updated and revised.

Chapter 4
1. Clinical Application box on gene therapy updated and new information added on genome editing with CRISPR.
2. New information added on pellagra in relation to niacin deficiency.

Chapter 5
1. Updated and expanded discussion of urea formation and oxidative deamination.
2. Lactate substituted for lactic acid when discussing the circulating form of the molecule.

Chapter 6
1. Explanation of facilitated diffusion updated and expanded.
2. Description of junctional complexes updated and expanded.
3. Osmosis discussion modified to include effects of relative volumes of solutions.

Chapter 7
1. Discussion of multiple sclerosis updated.
2. Updated information about regeneration of axons in the CNS.
3. Updated and expanded discussion of functions of astrocytes with new description of neuron-glial crosstalk.
4. Updated discussion of serotonin neurotransmission and the action of SSRIs.
5. Updated discussion of the physiological roles of endocannabinoids, and updated discussion of the untoward effects of marijuana on the brain.
6. Discussion of purinergic neurotransmission updated and expanded.

Chapter 8
1. Updated and expanded discussion of CSF formation and circulation, and the different layers of meninges.
2. Updated and expanded discussion of the amygdala and its functions.
3. Updated and expanded discussion of dendritic spines and their functions.
4. Updated and expanded discussion of the neurobiological basis of addiction.
5. Updated and expanded discussion of the cerebellum and Purkinje cells.
6. Table of ascending tracts of the spinal cord (table 8.4) updated.

Chapter 10
1. Updated and expanded discussion of touch and pressure receptors in the skin.
2. Discussion of neural pathways from cutaneous receptors updated and expanded.
3. Discussion of the physiology of the basilar membrane and spiral organ updated and expanded.
4. Discussion of retinitis pigmentosa updated.
5. Updated discussion of retinal ganglion cells.

Chapter 11
1. Updated discussion of the action of steroid hormones.
2. Discussion of the role of adrenocortical steroids in the stress response updated and expanded.
3. Discussion of insulin processing updated and expanded, with inclusion of C-protein.
4. Updated discussion of the regulation of the suprachiasmatic nucleus in circadian rhythms.

Chapter 12
1. Motor end plate description expanded and updated.
2. Discussion of the motor end plate updated and expanded.
3. Cross-bridge cycle description updated and revised.
4. Revised and updated discussion of motor units and recruitment in muscle contraction.
5. Description of muscle fiber types and effects of exercise updated and expanded.
6. Updated and expanded discussion of the health benefits of exercise and myokines released by skeletal muscles.
7. Discussion of satellite cells and sarcopenia updated and expanded.
8. Updated discussion of intercalated discs in cardiac muscle.

Chapter 13
1. Updated discussion of ABO system antibodies.
2. Updated discussion of platelet function.
3. Discussion of fibrillation updated and expanded.
4. Discussion of atrial fibrillation updated and expanded.

Chapter 14

1. Revised discussion of the relationship between stroke volume and peripheral resistance.
2. Updated and expanded discussion of how osmoreceptors stimulate thirst.
3. Discussion of the relationship between salt intake and blood pressure updated and revised.
4. New discussion of muscle vascularity and its significance.
5. Updated and expanded discussions of the regulation of blood flow to the abdominal organs and to the brain during exercise.
6. Updated and expanded discussion of the cardiovascular effects of endurance training.
7. Updated and expanded discussion of the effects of regular exercise on resting heart rate, blood pressure, and blood volume.
8. Discussion of the regulation of cutaneous blood flow updated and expanded.
9. Updated discussion of the baroreceptor reflex.
10. Updated and expanded discussion of the physiological factors that contribute to essential hypertension.
11. Discussion of congestive heart failure updated.

Chapter 15

1. New discussion of "trained immunity" in the innate immune system.
2. Revised discussion of fixed phagocytes and new discussion of the inflammasome concept.
3. Discussion of leukocyte migration to infected site updated and expanded.
4. New discussion of how nucleic acids from microbes stimulate inflammation through the intermediary action of STING ("stimulator of interferon genes").
5. Updated and expanded discussion of mast cell molecules in a local inflammation.
6. New discussion of VDJ recombination in the production of diverse antibodies.
7. Updated discussion of possible future methods to combat AIDS.
8. Updated discussion of tumor biology and immunology.
9. New discussion of innate lymphoid cells and noncytotoxic innate helper lymphoid cells.
10. Updated discussion of natural killer cells.
11. New discussion of immune checkpoint blockade therapy against tumors.

Chapter 16

1. New discussion of mucociliary clearance and its relationship to cystic fibrosis and smoking.
2. New discussion of measuring pulmonary pressures in centimeters of water instead of millimeters of mercury and how the pressures change during breathing.
3. Updated and expanded discussion of the functional residual capacity.
4. Updated discussion of asthma and COPD.
5. Discussion of the structure and function of the carotid bodies updated and expanded.

6. Updated and expanded discussion of the roles of pulmonary stretch receptors in the Hering-Breuer reflex and in control of normal tidal volume breathing.
7. Updated discussion of thalassemia.
8. Distinction between respiratory and metabolic components of acid-base balance revised.
9. Updated and expanded discussion of the hypoxic ventilator response.
10. Updated and expanded discussion of the effects of hypoxia on the pulmonary circulation.
11. Updated discussion of the effects of high altitude on the red blood cell count.

Chapter 17

1. Updated and expanded explanation of the mechanisms of tubuloglomerular feedback.
2. New explanation of the role of the primary cilium in regulating fluid flow in the distal nephron.
3. Updated and expanded discussion of the different functions of arginine vasopressin, including its function as the antidiuretic hormone.
4. Discussion of the effect of ADH on urine concentration updated and expanded.
5. New description of multispecific transporters in the renal nephron.
6. New discussion of the role of Na^+-Cl^- cotransporters in the distal convoluted tubule and their targeting by thiazide diuretics.
7. Updated and expanded discussion of how the nephron regulates homeostasis of the plasma K^+ concentration.
8. Updated and expanded discussion of how thiazide diuretics affect Na^+ and K^+ in the tubular fluid.
9. Updated and expanded discussion of the functions of the H^+/K^+ pumps in the nephron.
10. Discussion of the role of renal reabsorption of bicarbonate in acid-base balance updated and expanded.
11. New discussion of diabetic kidney disease and expanded description of renal insufficiency.

Chapter 18

1. Updated and expanded discussion of bariatric surgeries.
2. Discussion of the specialized cells at the bottom of the intestinal crypts updated.
3. Discussion of the nature and numbers of the intestinal microbiota updated.
4. New discussion of the relationship between obesity and the intestinal microbiota.
5. New discussion of the importance of dietary fiber on short chains fatty acid production and how this affects health.
6. New discussion of the different physiological roles of intestinal mucus.
7. Updated and expanded discussion of the importance of intestinal microbiota diversity to health.
8. Discussion of the different causes of diarrhea updated and expanded.

New to This Edition

9. New discussion of nonalcoholic steatohepatitis (NASH).
10. New distinction made between the metabolic fate of short and long chain fatty acids produced by the intestinal microbiota.

Chapter 19

1. New discussion of the dangers of too much antioxidants.
2. Updated and expanded discussion of visceral fat.
3. New discussion of adiponectin and insulin resistance.
4. New discussion of the role of motilin in hunger.
5. Updated and expanded discussion of the orexigenic and anorexigenic neurons in the arcuate nucleus and the control of hunger.
6. Updated discussion of the physiological effects of leptin.
7. New discussion of beige adipocytes and the "browning" of white fat.
8. Updated discussion of the physiological actions of brown and beige fat and their relation to obesity.
9. Updated discussion of the treatments for type 1 diabetes mellitus.
10. Updated discussion of the causes of hyperglycemia in type 2 diabetes mellitus.
11. Discussion of the hormonal regulation of bone updated and expanded.

Chapter 20

1. Updated and expanded discussion of genomic imprinting.
2. Updated discussion of regulation of puberty and menstruation by leptin.
3. Discussion of the regulation of the onset of puberty in boys updated and expanded.
4. Updated discussion of spermatogenesis along the seminiferous tubules.
5. Regulation of semen coagulation and liquefaction updated.
6. New discussion of developing male contraceptive possibilities.
7. New discussion of the systemic effects of estradiol.
8. New Exercise Application box on the Female Athlete Triad.
9. Updated discussion of menopause with new discussion of perimenopause.
10. Discussion of sperm flagellar movements during hyperactivation updated.
11. Updated and expanded discussion of circulation within the placenta.
12. Updated and expanded discussion of lactation and its hormonal control.

50% of the country's students are not ready for A&P

LearnSmart® Prep can help!

Improve preparation for the course and increase student success with the only adaptive Prep tool available for students today. Areas of individual weaknesses are identified in order to help students improve their understanding of core course areas needed to succeed.

LEARNSMART®
Prep for A&P

Students seek lab time that fits their busy schedules. Anatomy & Physiology REVEALED 3.2, our Virtual Dissection tool, allows students to practice anytime, anywhere, and now features enhanced physiology interactives with clinical and 3D animations.

Anatomy & Physiology | REVEALED® 3.2
Virtual dissection

Bringing to life complex processes is a challenge. Ph.I.L.S. 4.0 is the perfect way to reinforce key physiology concepts with powerful lab experiments. Tools like physiology interactives, Ph.I.L.S., and world-class animations make it easier than ever.

Ph.I.L.S.
Physiology supplements

The Practice Atlas for Anatomy & Physiology is a new interactive tool that pairs images of common anatomical models with stunning cadaver photography. This atlas allows students to practice naming structures on both models and human bodies, anytime and anywhere.

Practice Atlas for Anatomy & Physiology

 **connect®**

McGraw-Hill Connect® is a highly reliable, easy-to-use homework and learning management solution that utilizes learning science and award-winning adaptive tools to improve student results.

Homework and Adaptive Learning

- Connect's assignments help students contextualize what they've learned through application, so they can better understand the material and think critically.
- Connect will create a personalized study path customized to individual student needs through SmartBook®.
- SmartBook helps students study more efficiently by delivering an interactive reading experience through adaptive highlighting and review.

Over **7 billion questions** have been answered, making McGraw-Hill Education products more intelligent, reliable, and precise.

Connect's Impact on Retention Rates, Pass Rates, and Average Exam Scores

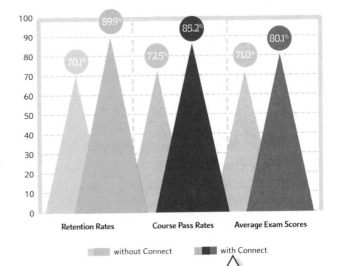

| | without Connect | with Connect |

Using **Connect** improves retention rates by **19.8%**, passing rates by **12.7%, and** exam scores by **9.1%.**

73% of instructors who use **Connect** require it; instructor satisfaction **increases** by 28% when **Connect** is required.

Quality Content and Learning Resources

- Connect content is authored by the world's best subject matter experts, and is available to your class through a simple and intuitive interface.
- The Connect eBook makes it easy for students to access their reading material on smartphones and tablets. They can study on the go and don't need internet access to use the eBook as a reference, with full functionality.
- Multimedia content such as videos, simulations, and games drive student engagement and critical thinking skills.

Robust Analytics and Reporting

©Hero Images/Getty Images

- Connect Insight® generates easy-to-read reports on individual students, the class as a whole, and on specific assignments.

- The Connect Insight dashboard delivers data on performance, study behavior, and effort. Instructors can quickly identify students who struggle and focus on material that the class has yet to master.

- Connect automatically grades assignments and quizzes, providing easy-to-read reports on individual and class performance.

Impact on Final Course Grade Distribution

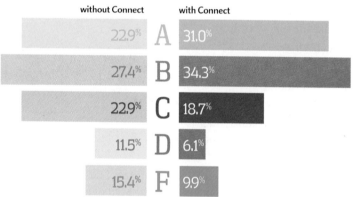

without Connect		with Connect
22.9%	A	31.0%
27.4%	B	34.3%
22.9%	C	18.7%
11.5%	D	6.1%
15.4%	F	9.9%

More students earn **As** and **Bs** when they use **Connect**.

Trusted Service and Support

- Connect integrates with your LMS to provide single sign-on and automatic syncing of grades. Integration with Blackboard®, D2L®, and Canvas also provides automatic syncing of the course calendar and assignment-level linking.

- Connect offers comprehensive service, support, and training throughout every phase of your implementation.

- If you're looking for some guidance on how to use Connect, or want to learn tips and tricks from super users, you can find tutorials as you work. Our Digital Faculty Consultants and Student Ambassadors offer insight into how to achieve the results you want with Connect.

Acknowledgments

REVIEWERS

Michael Ballentine
Kansas City Kansas Community College

Erwin Bautista
University of California—Davis

Yerko Berrocal
Florida International University

Subha Bhaskaran
Oakland University

Gerrit J. Bouma
Colorado State University

Tyson Chappell
Utah State University—Eastern

Raymond J. Clark
MiraCosta College

Corey Cleland
James Madison University

Scott Crousillac
Louisiana State University

Robert Farrell
Penn State University

Michael S. Finkler
Indiana University Kokomo

Shawn Flanagan
University of Iowa

Todd Gordon
Kansas City Kansas Community College

Jessica Habashi
Utah State University—Brigham City

Cindy L. Hansen
Community College of Rhode Island

Valerie L. Hedges
Northern Michigan University

Kelly Johnson
University of Kansas

Warren R. Jones
Loyola University Chicago

Beth Kersten
State College of Florida—Sarasota

Kevin Langford
Stephen F. Austin University

Alexander Makarov
Johnson County Community College

Nick Ritucci
Wright State University

Jennifer Rogers
The University of Iowa

Kyla Ross
Georgia State University

Bryan Rourke
California State University, Long Beach

Amber D. Ruskell-Lamer
Southeastern Community College

Otto A Sanchez
University of Minnesota Division of Renal Disease and
Hypertension Department of Internal Medicine

Merideth Sellars
Columbus State Community College

Nicole Shives
Iowa Western Community College—Council Bluffs

Todd Shoepe
Loyola Marymount University

Maureen Walter
Florida International University—BBC

Karri Haen Whitmer
Iowa State University

Contents

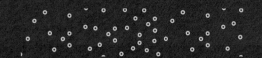

1

The Study of Body Function

As you study the sections of this chapter, you can see how your new knowledge can be applied to interesting health issues that may be important to know in your future career as a health professional. This can add zest to your studies and increase your motivation to truly understand physiological concepts, rather than to simply memorize facts for examinations. Each chapter begins with a medical mystery for you to solve, using information in the text of that chapter and "Clinical Investigation Clues" within the chapter.

For example, suppose Linda goes for a medical examination where her body temperature is measured, and she gives a fasting blood sample to test for glucose. Your first Clinical Investigation challenge is to determine the medical significance of these physiological tests.

CHAPTER OUTLINE

1.1 INTRODUCTION TO PHYSIOLOGY

Human physiology is the study of how the human body functions, with emphasis on specific cause-and-effect mechanisms. Knowledge of these mechanisms has been obtained experimentally through applications of the scientific method.

LEARNING OUTCOMES

After studying this section, you should be able to:

1. Describe the scientific study of human physiology.
2. Describe the characteristics of the scientific method.

Physiology (from the Greek *physis* = nature; *logos* = study) is the study of biological function—of how the body works, from molecular mechanisms within cells to the actions of tissues, organs, and systems, and how the organism as a whole accomplishes particular tasks essential for life. In the study of physiology, the emphasis is on mechanisms—with questions that begin with the word *how* and answers that involve cause-and-effect sequences. These sequences can be woven into larger and larger stories that include descriptions of the structures involved (anatomy) and that overlap with the sciences of chemistry and physics.

The separate facts and relationships of these cause-and-effect sequences are derived empirically from experimental evidence. Explanations that seem logical are not necessarily true; they are only as valid as the data on which they are based, and they can change as new techniques are developed and further experiments are performed. The ultimate objective of physiological research is to understand the normal functioning of cells, organs, and systems. A related science—*pathophysiology*—is concerned with how physiological processes are altered in disease or injury.

Pathophysiology and the study of normal physiology complement one another. For example, a standard technique for investigating the functioning of an organ is to observe what happens when the organ is surgically removed from an experimental animal or when its function is altered in a specific way. This study is often aided by "experiments of nature"—diseases—that involve specific damage to the functioning of an organ. The study of disease processes has thus aided our understanding of normal functioning, and the study of normal physiology has provided much of the scientific basis of modern medicine. This relationship is recognized by the Nobel Prize committee, whose members award prizes in the category "Physiology or Medicine."

The physiology of invertebrates and of different vertebrate groups is studied in the science of *comparative physiology*. Much of the knowledge gained from comparative physiology has benefited the study of human physiology. This is because animals, including humans, are more alike than they are different. This is especially true when comparing humans with other mammals. The small differences in physiology between humans and other mammals can be of crucial importance in the development of pharmaceutical drugs (discussed later in this section), but these differences are relatively slight in the overall study of physiology.

Scientific Method

All of the information in this text has been gained by people applying the **scientific method.** Although many different techniques are involved when people apply the scientific method, all share three attributes: (1) confidence that the natural world, including ourselves, is ultimately explainable in terms we can understand; (2) descriptions and explanations of the natural world that are honestly based on observations and that could be modified or refuted by other observations; and (3) humility, or the willingness to accept the fact that we could be wrong. If further study should yield conclusions that refuted all or part of an idea, the idea would have to be modified accordingly. In short, the scientific method is based on a confidence in our rational ability, honesty, and humility. Practicing scientists may not always display these attributes, but the validity of the large body of scientific knowledge that has been accumulated—as shown by the technological applications and the predictive value of scientific hypotheses—is ample testimony to the fact that the scientific method works.

The scientific method involves specific steps. After certain observations regarding the natural world are made, a **hypothesis** is formulated. In order for this hypothesis to be scientific, it must be capable of being refuted by experiments or other observations of the natural world. For example, one might hypothesize that people who exercise regularly have a lower resting pulse rate than other people. Experiments are conducted, or other observations are made, and the results are analyzed. Conclusions are then drawn as to whether the new data either refute or support the hypothesis. If the hypothesis survives such testing, it might be incorporated into a more general **theory.** Scientific theories are thus not simply conjectures; they are statements about the natural world that incorporate a number of proven hypotheses. They serve as a logical framework by which these hypotheses can be interrelated and provide the basis for predictions that may as yet be untested.

The hypothesis in the preceding example is scientific because it is *testable;* the pulse rates of 100 athletes and 100 sedentary people could be measured, for example, to see if there were statistically significant differences. If there were, the statement that athletes, on the average, have lower resting pulse rates than other people would be justified *based on these data.* One must still be open to the fact that this conclusion could be wrong. Before the discovery could become generally accepted as fact, other scientists would have to consistently replicate the results. Scientific theories are based on *reproducible* data.

It is quite possible that when others attempt to replicate the experiment, their results will be slightly different. They may then construct scientific hypotheses that the differences in resting pulse rate also depend on other factors, such as the nature of the exercise performed. When scientists attempt to test these hypotheses, they will likely encounter new problems requiring new explanatory hypotheses, which then must be tested by additional experiments.

In this way, a large body of highly specialized information is gradually accumulated, and a more generalized explanation (a scientific theory) can be formulated. This explanation will almost always be different from preconceived notions. People who follow the scientific method will then appropriately modify their concepts, realizing that their new ideas will probably have to be changed again in the future as additional experiments are performed.

Use of Measurements, Controls, and Statistics

Suppose you wanted to test the hypothesis that a regular exercise program causes people to have a lower resting heart rate. First, you would have to decide on the nature of the exercise program. Then, you would have to decide how the heart rate (or pulse rate) would be measured. This is a typical problem in physiology research because the testing of most physiological hypotheses requires quantitative **measurements.**

The group that is subject to the testing condition—in this case, exercise—is called the **experimental group.** A measurement of the heart rate for this group would be meaningful only if it is compared to that of another group, known as the **control group.** How shall this control group be chosen? Perhaps the subjects could serve as their own controls—that is, a person's resting heart rate could be measured before and after the exercise regimen. If this isn't possible, a control group could be other people who do not follow the exercise program. The choice of control groups is often a controversial aspect of physiology studies. In this example, did the people in the control group really refrain from *any* exercise? Were they comparable to the people in the experimental group with regard to age, sex, ethnicity, body weight, health status, and so on? You can see how difficult it could be in practice to get a control group that could satisfy any potential criticism.

Another possible criticism could be bias in the way that the scientists perform the measurements. This bias could be completely unintentional; scientists are human, after all, and they may have invested months or years in this project. To prevent such bias, the person doing the measurements often does not know if a subject is part of the experimental or the control group. This is known as a *blind measurement.*

Now suppose the data are in and it looks like the experimental group indeed has a lower average resting heart rate than the control group. But there is overlap—some people in the control group have measurements that are lower than some people in the experimental group. Is the difference in the average measurements of the groups due to a real physiological difference, or is it due to chance variations in the measurements? Scientists attempt to test the *null hypothesis* (the hypothesis that the difference is due to chance) by employing the mathematical tools of **statistics.** If the statistical results so warrant, the null hypothesis can be rejected and the experimental hypothesis can be deemed to be supported by this study.

The statistical test chosen will depend upon the design of the experiment, and it can also be a source of contention among scientists in evaluating the validity of the results. Because of the nature of the scientific method, "proof" in science is always provisional. Some other researchers, employing the scientific method in a different way (with different measuring techniques, experimental procedures, choice of control groups, statistical tests, and so on), may later obtain different results. The scientific method is thus an ongoing enterprise.

The results of the scientific enterprise are written up as research articles, and these must be reviewed by other scientists who work in the same field before they can be published in **peer-reviewed journals.** More often than not, the reviewers will suggest that certain changes be made in the articles before they can be accepted for publication.

Examples of such peer-reviewed journals that publish articles in many scientific fields include *Science* (www.sciencemag.org/), *Nature* (www.nature.com/nature/), and *Proceedings of the National Academy of Sciences* (www.pnas.org/). Review articles on physiology can be found in *Annual Review of Physiology* (physiol.annualreviews.org/), *Physiological Reviews* (physrev.physiology.org/), and *Physiology* (physiologyonline.physiology.org). Medical research journals, such as the *New England Journal of Medicine* (content.nejm.org/) and *Nature Medicine* (www.nature.com/nm/), also publish articles of physiological interest. There are also many specialty journals in areas of physiology such as neurophysiology, endocrinology, and cardiovascular physiology.

Students who wish to look online for scientific articles published in peer-reviewed journals that relate to a particular subject can do so at the National Library of Medicine website, *PubMed* (www.ncbi.nlm.nih.gov/entrez/query.fcgi).

Development of Pharmaceutical Drugs

The development of new pharmaceutical drugs can serve as an example of how the scientific method is used in physiology and its health applications. The process usually starts with basic physiological research, often at cellular and molecular levels. Perhaps a new family of drugs is developed using cells in tissue culture (*in vitro,* or outside the body). For example, cell physiologists studying membrane transport may discover that a particular family of compounds blocks membrane channels for calcium ions (Ca^{2+}). Because of their knowledge of physiology, other scientists may predict that a drug of this nature might be useful in the treatment of hypertension (high blood pressure). This drug may then be tried in animal experiments.

If a drug is effective at extremely low concentrations *in vitro* (in cells cultured outside of the body), there is a chance that it may work *in vivo* (in the body) at concentrations low enough not to be toxic (poisonous). This possibility must be thoroughly tested utilizing experimental animals, primarily rats and mice. More than 90% of drugs tested in experimental animals are too toxic for further development. Only in those rare cases when the toxicity is low enough may development progress to human/clinical trials.

Biomedical research is often aided by **animal models** of particular diseases. These are strains of laboratory rats and mice that are genetically susceptible to particular diseases that resemble human diseases. Research utilizing laboratory animals typically takes several years and always precedes human (clinical) trials of promising drugs. It should be noted that this length of time does not include all of the years of "basic" physiological research (involving laboratory animals) that provided the scientific foundation for the specific medical application.

In **phase I clinical trials,** the drug is tested on healthy human volunteers. This is done to test its toxicity in humans and to study how the drug is "handled" by the body: how it is metabolized, how rapidly it is removed from the blood by the liver and kidneys, how it can be most effectively administered, and so on. If significant toxic effects are not observed, the drug can proceed to the next stage. In **phase II clinical trials,** the drug is tested on the target human population (for example, those with hypertension). Only in those exceptional cases where the drug seems to be effective but has minimal toxicity does testing move to the next phase. **Phase III trials** occur in many research centers across the country to maximize the number of test participants. At this point, the test population must include a sufficient number of subjects of both sexes, as well as people of different ethnic groups. In addition, people are tested who have other health problems besides the one that the drug is intended to benefit. For example, those who have diabetes in addition to hypertension would be included in this phase. If the drug passes phase III trials, it goes to the Food and Drug Administration (FDA) for approval. **Phase IV trials** test other potential uses of the drug. These "post-marketing studies" often reveal problems with the drug that were not previously evident.

Less than 10% of the tested drugs make it all the way through clinical trials to eventually become approved and marketed. This low success rate does not count those that fail after approval because of unexpected toxicity, nor does it take into account the great amount of drugs that fail earlier in research before clinical trials begin. Notice the crucial role of basic research, using experimental animals, in this process. Virtually every prescription drug on the market owes its existence to such research.

CHECKPOINTS

1. How has the study of physiology aided, and been aided by, the study of diseases?
2a. Describe the steps involved in the scientific method. What would qualify a statement as unscientific?
2b. Describe the different types of trials a new drug must undergo before it is "ready for market."

1.2 HOMEOSTASIS AND FEEDBACK CONTROL

The regulatory mechanisms of the body can be understood in terms of a single shared function: that of maintaining constancy of the internal environment. A state of relative constancy of the internal environment is known as homeostasis, maintained by negative feedback loops.

LEARNING OUTCOMES

After studying this section, you should be able to:

3. Define homeostasis, and identify the components of negative feedback loops.
4. Explain the role of antagonistic effectors in maintaining homeostasis, and the nature of positive feedback loops.
5. Give examples of how negative feedback loops involving the nervous and endocrine systems help to maintain homeostasis.

History of Physiology

The Greek philosopher Aristotle (384–322 B.C.) speculated on the function of the human body, but another ancient Greek, Erasistratus (304–250? B.C.), is considered to be the first to study physiology because he attempted to apply physical laws to understand human function. Galen (A.D. 130–201) wrote widely on the subject and was considered the supreme authority until the Renaissance. Physiology became a fully experimental science with the revolutionary work of the English physician William Harvey (1578–1657), who demonstrated that the heart pumps blood through a closed system of vessels.

However, the originator of modern physiology is the French physiologist Claude Bernard (1813–1878), who observed that the *milieu intérieur* (internal environment) remains remarkably constant despite changing conditions in the external environment. In a book entitled *The Wisdom of the Body,* published in 1932, the American physiologist Walter Cannon (1871–1945) coined the term **homeostasis** to describe this internal constancy. Cannon further suggested that the many mechanisms of physiological regulation have but one purpose—the maintenance of internal constancy. In the early 1950s, James Hardin extended Cannon's concept by proposing that homeostatic mechanisms maintain each physiological variable within a normal range by comparing its value to a desired, or *set point,* value (as will be described shortly).

Most of our present knowledge of human physiology has been gained in the twentieth century. However, new knowledge in the twenty-first century is being added at an ever more rapid pace, fueled in more recent decades by the revolutionary growth of molecular genetics and its associated biotechnologies, and by the availability of more powerful computers and other equipment. A very brief history of twentieth- and twenty-first-century physiology, limited by space to only two citations per decade, is provided in table 1.1.

Most of the citations in table 1.1 indicate the winners of Nobel Prizes. The **Nobel Prize in Physiology or Medicine** (a single prize category) was first awarded in 1901 to Emil Adolf von Behring, a pioneer in immunology who coined the term *antibody* and whose many other discoveries included the use of serum (containing antibodies) to treat diphtheria. Many scientists who might deserve a Nobel Prize never receive one, and the prizes are given for particular achievements and not others (Einstein didn't win his Nobel Prize in Physics for relativity, for example) and are often awarded many years after the discoveries were made. Nevertheless, the awarding of the Nobel Prize in Physiology or Medicine each year is a celebrated event in the biomedical community, and the awards can be a useful yardstick for tracking the course of physiological research over time.

Negative Feedback Loops

The concept of homeostasis has been of immense value in the study of physiology because it allows diverse regulatory mechanisms to be understood in terms of their "why" as well as their "how." The concept of homeostasis also provides a major foundation for medical diagnostic procedures. When a particular measurement of the internal environment, such as a blood measurement (table 1.2), deviates significantly from the normal range of values, it can be concluded that homeostasis is not being maintained and that the person is sick. A number of such measurements, combined with clinical observations, may allow the particular defective mechanism to be identified.

TABLE 1.1 | History of Twentieth- and Twenty-First-Century Physiology (two citations per decade)

1900	Karl Landsteiner discovers the A, B, and O blood groups.
1904	Ivan Pavlov wins the Nobel Prize for his work on the physiology of digestion.
1910	Sir Henry Dale describes properties of histamine.
1918	Earnest Starling describes how the force of the heart's contraction relates to the amount of blood in it.
1921	John Langley describes the functions of the autonomic nervous system.
1923	Sir Frederick Banting, Charles Best, and John Macleod win the Nobel Prize for the discovery of insulin.
1932	Sir Charles Sherrington and Lord Edgar Adrian win the Nobel Prize for discoveries related to the functions of neurons.
1936	Sir Henry Dale and Otto Loewi win the Nobel Prize for the discovery of acetylcholine in synaptic transmission.
1939–47	Albert von Szent-Györgyi explains the role of ATP and contributes to the understanding of actin and myosin in muscle contraction.
1949	Hans Selye discovers the common physiological responses to stress.
1953	Sir Hans Krebs wins the Nobel Prize for his discovery of the citric acid cycle.
1954	Hugh Huxley, Jean Hanson, R. Niedergerde, and Andrew Huxley propose the sliding filament theory of muscle contraction.
1962	Francis Crick, James Watson, and Maurice Wilkins win the Nobel Prize for determining the structure of DNA.
1963	Sir John Eccles, Sir Alan Hodgkin, and Sir Andrew Huxley win the Nobel Prize for their discoveries relating to the nerve impulse.
1971	Earl Sutherland wins the Nobel Prize for his discovery of the mechanism of hormone action.
1977	Roger Guillemin and Andrew Schally win the Nobel Prize for discoveries of the brain's production of peptide hormone.
1981	Roger Sperry wins the Nobel Prize for his discoveries regarding the specializations of the right and left cerebral hemispheres.
1986	Stanley Cohen and Rita Levi-Montalcini win the Nobel Prize for their discoveries of growth factors regulating the nervous system.
1994	Alfred Gilman and Martin Rodbell win the Nobel Prize for their discovery of the functions of G-proteins in signal transduction in cells.
1998	Robert Furchgott, Louis Ignarro, and Ferid Murad win the Nobel Prize for discovering the role of nitric oxide as a signaling molecule in the cardiovascular system.
2004	Linda B. Buck and Richard Axel win the Nobel Prize for their discoveries of odorant receptors and the organization of the olfactory system.
2006	Andrew Z. Fine and Craig C. Mello win the Noble Prize for their discovery of RNA interference by short, double-stranded RNA molecules.
2012	Sir John Gurdon and Shinya Yamanaka win the Nobel Prize for their discovery that mature cells can be reprogrammed to become pluripotent (like embryonic cells).

TABLE 1.2 | Approximate Normal Ranges for Measurements of Some Fasting Blood Values

Measurement	Normal Range
Arterial pH	7.35–7.45
Bicarbonate	24–28 mEq/L
Sodium	135–145 mEq/L
Calcium	4.5–5.5 mEq/L
Oxygen content	17.2–22.0 ml/100 ml
Urea	12–35 mg/100 ml
Amino acids	3.3–5.1 mg/100 ml
Protein	6.5–8.0 g/100 ml
Total lipids	400–800 mg/100 ml
Glucose	70–99 mg/100 ml

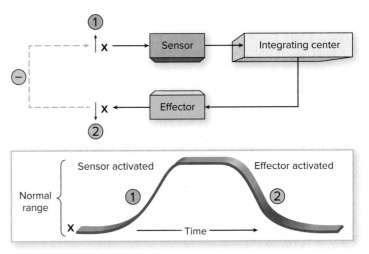

FIGURE 1.1   A rise in some factor of the internal environment (↑X) is detected by a sensor. This information is relayed to an integrating center, which causes an effector to produce a change (1) in the opposite direction (↓X). The initial deviation is thus reversed (2), completing a negative feedback loop (shown by the dashed arrow and negative sign). The numbers indicate the sequence of changes.

In order for internal constancy to be maintained, changes in the body must stimulate *receptors,* which function as **sensors** that can send information to an **integrating center.** This allows the integrating center to detect changes from a **set point.** The set point is analogous to the temperature set on a house thermostat. In a similar manner, there is a set point for body temperature, blood glucose concentration, the tension on a tendon, and so on. The integrating center is often a particular region of the brain or spinal cord, but it can also be a group of cells in an endocrine gland. A number of different sensors may send information to a particular integrating center, which can then integrate this information and direct the responses of **effectors**—generally muscles or glands. The integrating center may cause increases or decreases in effector action to counter the deviations from the set point and defend homeostasis.

The thermostat of a house can serve as a simple example. Suppose you set the thermostat at a set point of 70° F. If the temperature in the house rises sufficiently above the set point, a sensor connected to an integrating center within the thermostat will detect that deviation and turn on the air conditioner (the effector in this example). The air conditioner will turn off when the room temperature falls and the thermostat no longer detects a deviation from the set-point temperature. However, this simple example gives a wrong impression: the effectors in the body are generally increased or decreased in activity, *not* just turned on or off. Because of this, negative feedback control in the body works far more efficiently than does a house thermostat.

If the body temperature exceeds the set point of 37° C, sensors in a part of the brain detect this deviation and, acting via an integrating center (also in the brain), stimulate activities of effectors (including sweat glands) that lower the temperature. For another example, if the blood glucose concentration falls below normal, the effectors act to increase the blood glucose. One can think of the effectors as "defending" the set points against deviations. Because the activity of the effectors is

influenced by the effects they produce, and because this regulation is in a negative, or reverse, direction, this type of control system is known as a **negative feedback loop** (fig. 1.1). (Notice that in figure 1.1 and in all subsequent figures, negative feedback is indicated by a dashed line and a negative sign.)

The nature of the negative feedback loop can be understood by again referring to the analogy of the thermostat and air conditioner. After the air conditioner has been on for some time, the room temperature may fall significantly below the set point of the thermostat. When this occurs, the air conditioner will be turned off. The effector (air conditioner) is turned on by a high temperature and, when activated, produces a negative change (lowering of the temperature) that ultimately causes the effector to be turned off. In this way, constancy is maintained.

It is important to realize that these negative feedback loops are continuous, ongoing processes. Thus, a particular nerve fiber that is part of an effector mechanism may always display some activity, and a particular hormone that is part of another effector mechanism may always be present in the blood. The nerve activity and hormone concentration may decrease in response to deviations of the internal environment in one direction (fig. 1.1), or they may increase in response to deviations in the opposite direction (fig. 1.2). Changes from the normal range in either direction are thus compensated for by reverse changes in effector activity.

Because negative feedback loops respond after deviations from the set point have stimulated sensors, the internal environment is never absolutely constant. Homeostasis is best conceived as a state of **dynamic constancy** in which conditions are stabilized above and below the set point. These conditions can be measured quantitatively, in degrees Celsius for body temperature, for example, or in milligrams per deciliter (one-tenth of a liter) for blood glucose. The set point can be taken as the average value within the normal range of measurements (fig. 1.3).

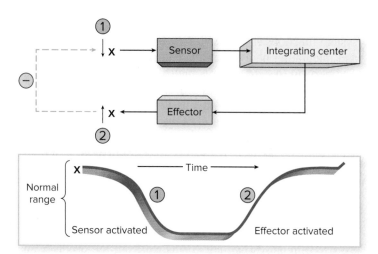

FIGURE 1.2   A fall in some factor of the internal environment (↓X) is detected by a sensor. (Compare this negative feedback loop with that shown in figure 1.1.)

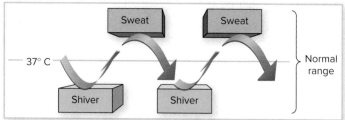

FIGURE 1.4   How body temperature is maintained within the normal range. The body temperature normally has a set point of 37° C. This is maintained, in part, by two antagonistic mechanisms—shivering and sweating. Shivering is induced when the body temperature falls too low, and it gradually subsides as the temperature rises. Sweating occurs when the body temperature is too high, and it diminishes as the body temperature falls. Most aspects of the internal environment are regulated by the antagonistic actions of different effector mechanisms.

See the *Test Your Quantitative Ability* section of the **Review Activities** at the end of this chapter.

FIGURE 1.3   Negative feedback loops maintain a state of dynamic constancy within the internal environment. The completion of the negative feedback loop is indicated by negative signs.

Antagonistic Effectors

Most factors in the internal environment are controlled by several effectors, which often have antagonistic actions. Control by antagonistic effectors is sometimes described as "push-pull," where the increasing activity of one effector is accompanied by decreasing activity of an antagonistic effector. This affords a finer degree of control than could be achieved by simply switching one effector on and off.

Room temperature can be maintained, for example, by simply turning an air conditioner on and off, or by just turning a heater on and off. A much more stable temperature, however, can be achieved if the air conditioner and heater are both controlled by a thermostat. Then the heater is turned on when the air conditioner is turned off, and vice versa. Normal body temperature is maintained about a set point of 37° C by the antagonistic effects of sweating, shivering, and other mechanisms (fig. 1.4).

The blood concentrations of glucose, calcium, and other substances are regulated by negative feedback loops involving hormones that promote opposite effects. Insulin, for example, lowers blood glucose, and other hormones raise the blood glucose concentration. The heart rate, similarly, is controlled by nerve fibers that produce opposite effects: stimulation of one group of nerve fibers increases heart rate; stimulation of another group slows the heart rate.

Quantitative Measurements

In order to study physiological mechanisms, scientists must measure specific values and mathematically determine such statistics as their normal range, their averages, and their deviations from the average (which can represent the set point). For these and other reasons, quantitative measurements are basic to the science of physiology. One example of this, and of the actions of antagonistic mechanisms in maintaining homeostasis, is shown in figure 1.5. Blood glucose concentrations were measured in five healthy people before and after an injection of insulin, a hormone that acts to lower the blood glucose concentration. A graph of the data reveals that the blood glucose concentration decreased rapidly but was brought back up to

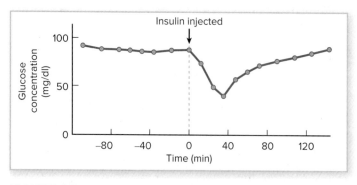

FIGURE 1.5   Homeostasis of the blood glucose concentration. Average blood glucose concentrations of five healthy individuals are graphed before and after a rapid intravenous injection of insulin. The "0" indicates the time of the injection. The blood glucose concentration is first lowered by the insulin injection, but is then raised back to the normal range (by hormones antagonistic to insulin that stimulate the liver to secrete glucose into the blood). Homeostasis of blood glucose is maintained by the antagonistic actions of insulin and several other hormones.

normal levels within 80 minutes after the injection. This demonstrates that negative feedback mechanisms acted to restore homeostasis in this experiment. These mechanisms involve the action of hormones whose effects are antagonistic to that of insulin—that is, they promote the secretion of glucose from the liver (see chapter 19).

Positive Feedback

Constancy of the internal environment is maintained by effectors that act to compensate for the change that served as the stimulus for their activation; in short, by negative feedback loops. A thermostat, for example, maintains a constant temperature by increasing heat production when it is cold and decreasing heat production when it is warm. The opposite occurs during **positive feedback**—in this case, the action of effectors *amplifies* those changes that stimulated the effectors. A thermostat that works by positive feedback, for example, would increase heat production in response to a rise in temperature.

It is clear that homeostasis must ultimately be maintained by negative rather than by positive feedback mechanisms. The effectiveness of some negative feedback loops, however, is increased by positive feedback mechanisms that amplify the actions of a negative feedback response. Blood clotting, for example, occurs as a result of a sequential activation of clotting factors; the activation of one clotting factor results in activation of many in a positive feedback cascade. In this way, a single change is amplified to produce a blood clot. Formation of the clot, however, can prevent further loss of blood, and thus represents the completion of a negative feedback loop that restores homeostasis.

Two other examples of positive feedback in the body are both related to the female reproductive system. One of these examples occurs when estrogen, secreted by the ovaries, stimulates the woman's pituitary gland to secrete LH (luteinizing hormone). This stimulatory, positive feedback effect creates an "LH surge" (very rapid rise in blood LH concentrations) that triggers ovulation. Interestingly, estrogen secretion after ovulation has an inhibitory, negative feedback, effect on LH secretion (this is the physiological basis for the birth control pill, discussed in chapter 20). Another example of positive feedback is contraction of the uterus during childbirth (parturition). Contraction of the uterus is stimulated by the pituitary hormone oxytocin, and the secretion of oxytocin is increased by sensory feedback from contractions of the uterus during labor. The strength of uterine contractions during labor is thus increased through positive feedback. The mechanisms involved in labor are discussed in more detail in chapter 20 (see fig. 20.50).

Neural and Endocrine Regulation

Homeostasis is maintained by two general categories of regulatory mechanisms: (1) those that are **intrinsic,** or "built into" the organs being regulated (such as molecules produced in the walls of blood vessels that cause vessel dilation or constriction); and (2) those that are **extrinsic,** as in regulation of an organ by the nervous and endocrine systems. The endocrine system functions closely with the nervous system in regulating and integrating body processes and maintaining homeostasis. The nervous system controls the secretion of many endocrine glands, and some hormones in turn affect the function of the nervous system. Together, the nervous and endocrine systems regulate the activities of most of the other systems of the body.

Regulation by the endocrine system is achieved by the secretion of chemical regulators called **hormones** into the blood, which carries the hormones to all organs in the body. Only specific organs can respond to a particular hormone, however; these are known as the **target organs** of that hormone.

Nerve fibers are said to *innervate* the organs that they regulate. When stimulated, these fibers produce electrochemical nerve impulses that are conducted from the origin of the fiber to its terminals in the target organ innervated by the fiber. These target organs can be muscles or glands that may function as effectors in the maintenance of homeostasis.

For example, we have negative feedback loops that help maintain homeostasis of arterial blood pressure, in part by adjusting the heart rate. If everything else is equal, blood pressure is lowered by a decreased heart rate and raised by an increased heart rate. This is accomplished by regulating the activity of the autonomic nervous system, as will be discussed in later chapters. Thus, a fall in blood pressure—produced daily as we go from a lying to a standing position—is compensated by a faster heart rate (fig. 1.6). As a consequence of this negative feedback loop, our heart rate varies as we go through our day, speeding up and slowing down, so that we can maintain homeostasis of blood pressure and keep it within limits.

Feedback Control of Hormone Secretion

The nature of the endocrine glands, the interaction of the nervous and endocrine systems, and the actions of hormones will be discussed in detail in later chapters. For now, it is sufficient to describe the regulation of hormone secretion very broadly, because it so superbly illustrates the principles of homeostasis and negative feedback regulation.

Hormones are secreted in response to specific chemical stimuli. A rise in the plasma glucose concentration, for example, stimulates insulin secretion from structures in the pancreas known as the *pancreatic islets*. Hormones are also secreted in response to nerve stimulation and stimulation by other hormones.

The secretion of a hormone can be inhibited by its own effects in a negative feedback manner. Insulin, as previously described, produces a lowering of blood glucose. Because a rise in blood glucose stimulates insulin secretion, a lowering of blood glucose caused by insulin's action inhibits further insulin secretion. This closed-loop control system is called **negative feedback inhibition** (fig. 1.7a).

Homeostasis of blood glucose is too important—the brain uses blood glucose as its primary source of energy—to entrust to the regulation of only one hormone, insulin. So, when blood

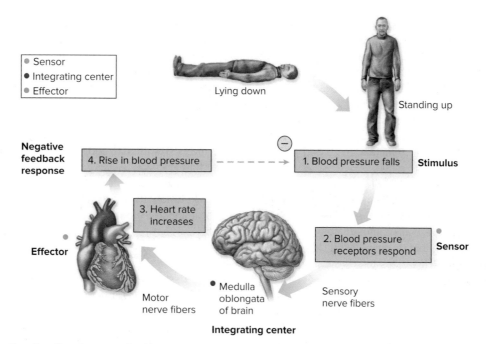

FIGURE 1.6 **Negative feedback control of blood pressure.** Blood pressure influences the activity of sensory neurons from the blood pressure receptors (sensors); a rise in pressure increases the firing rate, and a fall in pressure decreases the firing rate of nerve impulses. When a person stands up from a lying-down position, the blood pressure momentarily falls. The resulting decreased firing rate of nerve impulses in sensory neurons affects the medulla oblongata of the brain (the integrating center). This causes the motor nerves to the heart (effector) to increase the heart rate, helping to raise the blood pressure.

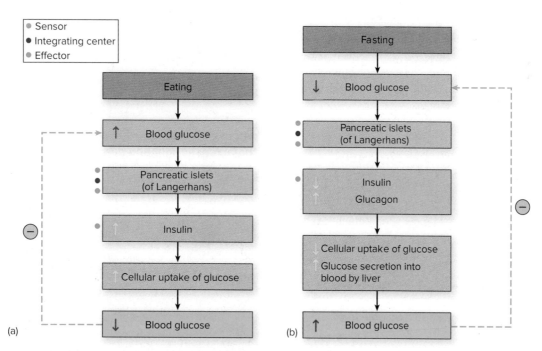

FIGURE 1.7 **Negative feedback control of blood glucose.** (*a*) The rise in blood glucose that occurs after eating carbohydrates is corrected by the action of insulin, which is secreted in increasing amounts at that time. (*b*) During fasting, when blood glucose falls, insulin secretion is inhibited and the secretion of an antagonistic hormone, glucagon, is increased. This stimulates the liver to secrete glucose into the blood, helping to prevent blood glucose from continuing to fall. In this way, blood glucose concentrations are maintained within a homeostatic range following eating and during fasting.

glucose falls during fasting, several mechanisms prevent it from falling too far (fig. 1.7*b*). First, insulin secretion decreases, preventing muscle, liver, and adipose cells from taking too much glucose from the blood. Second, the secretion of a hormone antagonistic to insulin, called *glucagon,* increases. Glucagon stimulates processes in the liver (breakdown of a stored, starchlike molecule called glycogen; chapter 2, section 2.2) that cause it to secrete glucose into the blood. Through these and other antagonistic negative feedback mechanisms, the blood glucose is maintained within a homeostatic range.

CLINICAL INVESTIGATION   CLUES

Clinical Investigation Clues are placed immediately following the text information that pertains to the Clinical Investigation for the chapter. Use these to solve the medical mystery—if you need to, re-read the information preceding the "Clues." You can check your answers against the Clinical Investigation Summaries at the end of the chapters. In this case, Linda had a normal resting body temperature and a normal fasting glucose concentration, suggesting that homeostasis of these values was being maintained.

CHECKPOINTS

3a. Define *homeostasis* and describe how this concept can be used to explain physiological control mechanisms.

3b. Define *negative feedback* and explain how it contributes to homeostasis. Illustrate this concept by drawing and labeling a negative feedback loop.

4. Describe *positive feedback* and explain how this process functions in the body.

5. Explain how the secretion of a hormone is controlled by negative feedback inhibition. Use the control of insulin secretion as an example.

1.3 THE PRIMARY TISSUES

The organs of the body are composed of four different primary tissues, each of which has its own characteristic structure and function. The activities and interactions of these tissues determine the physiology of the organs.

LEARNING OUTCOMES

After studying this section, you should be able to:

6. Distinguish the primary tissues and their subtypes.

7. Relate the structure of the primary tissues to their functions.

Although physiology is the study of function, it is difficult to properly understand the function of the body without some knowledge of its anatomy, particularly at a microscopic level.

Microscopic anatomy constitutes a field of study known as *histology.* The anatomy and histology of specific organs will be discussed together with their functions in later chapters. In this section, the common "fabric" of all organs is described.

Cells are the basic units of structure and function in the body. Cells that have similar functions are grouped into categories called *tissues.* The entire body is composed of only four major types of tissues. These **primary tissues** are (1) muscle, (2) nerve, (3) epithelial, and (4) connective tissues. Groupings of these four primary tissues into anatomical and functional units are called **organs.** Organs, in turn, may be grouped together by common functions into **systems.** The systems of the body act in a coordinated fashion to maintain the entire organism.

Muscle Tissue

Muscle tissue is specialized for contraction. There are three types of muscle tissue: **skeletal, cardiac,** and **smooth.** Skeletal muscle is often called *voluntary muscle* because its contraction is consciously controlled. Both skeletal and cardiac muscle tissues are **striated muscle tissue;** they have striations, or stripes, that extend across the width of the muscle cell (figs. 1.8 and 1.9). These striations are produced by a characteristic

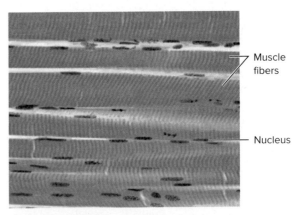

Muscle fibers

Nucleus

FIGURE 1.8   Skeletal muscle fibers showing the characteristic light and dark cross striations. Because of this feature, skeletal muscle is also called striated muscle tissue. **AP|R**
© McGraw-Hill Education/Al Telser, photographer

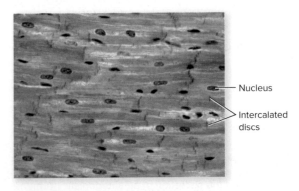

Nucleus

Intercalated discs

FIGURE 1.9   Human cardiac muscle. Notice the striated appearance and dark-staining intercalated discs. **AP|R**
© McGraw-Hill Education/Al Telser, photographer

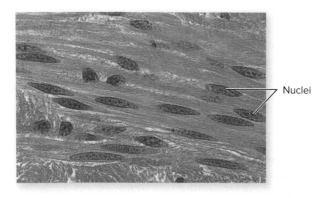

FIGURE 1.10   A photomicrograph of smooth muscle cells. Notice that these cells contain single, centrally located nuclei and lack striations. **AP|R** © McGraw-Hill Education/Dennis Strete, photographer

arrangement of contractile proteins, and for this reason skeletal and cardiac muscle have similar mechanisms of contraction. Smooth muscle (fig. 1.10) lacks these striations and has a different mechanism of contraction.

Skeletal Muscle

Skeletal muscles are generally attached to bones at both ends by means of tendons; hence, contraction produces movements of the skeleton. There are exceptions to this pattern, however. The tongue, superior portion of the esophagus, anal sphincter, and diaphragm are also composed of skeletal muscle, but they do not cause movements of the skeleton.

Beginning at about the fourth week of embryonic development, separate cells called *myoblasts* fuse together to form **skeletal muscle fibers,** or **myofiber** (from the Greek *myos* = muscle). Although myofibers are often referred to as skeletal muscle cells, each is actually a *syncytium,* or multinucleate mass formed from the union of separate cells. Despite their unique origin and structure, each myofiber contains mitochondria and other organelles (described in chapter 3) common to all cells.

The muscle fibers within a skeletal muscle are arranged in bundles, and within these bundles the fibers extend in parallel from one end of the bundle to the other. The parallel arrangement of muscle fibers (fig. 1.8) allows each fiber to be controlled individually: one can thus contract fewer or more muscle fibers and, in this way, vary the strength of contraction of the whole muscle. The ability to vary, or "grade," the strength of skeletal muscle contraction is needed for precise control of skeletal movements.

Cardiac Muscle

Although cardiac muscle is striated muscle tissue, it differs markedly from skeletal muscle in appearance. Cardiac muscle is found only in the heart, where the heart muscle cells, or **cardiac muscle (myocardial) cells,** are short, branched, and intimately interconnected to form a continuous fabric. Special areas of contact between adjacent cells stain darkly to show *intercalated discs* (fig. 1.9), which are characteristic of heart muscle.

The intercalated discs couple cardiac muscle cells together mechanically and electrically. Unlike skeletal muscles, therefore, the heart cannot produce a graded contraction by varying the number of cells stimulated to contract. Because of the way the heart is constructed, the stimulation of one cardiac muscle cell results in the stimulation of all other cells in the mass and a "wholehearted" contraction.

Smooth Muscle

As implied by the name, smooth muscle cells (fig. 1.10) do not have the striations characteristic of skeletal and cardiac muscle. Smooth muscle is found in the digestive tract, blood vessels, bronchioles (small air passages in the lungs), and the ducts of the urinary and reproductive systems. Circular arrangements of smooth muscle in these organs produce constriction of the *lumen* (cavity) when the muscle cells contract. The digestive tract also contains longitudinally arranged layers of smooth muscle. *Peristalsis* is the coordinated wavelike contractions of the circular and longitudinal smooth muscle layers that push food from the oral to the anal end of the digestive tract.

The three types of muscle tissue are discussed further in chapter 12.

Nerve Tissue

Nerve tissue consists of nerve cells, or **neurons,** which are specialized for the generation and conduction of electrical events, and **neuroglial (or glial) cells**, which compose the **neuroglia**. Neuroglia provide the neurons with structural support and perform a variety of functions that are needed for the normal physiology of the nervous system.

Each neuron consists of three parts: (1) a *cell body,* (2) *dendrites,* and (3) an *axon* (fig. 1.11). The cell body contains

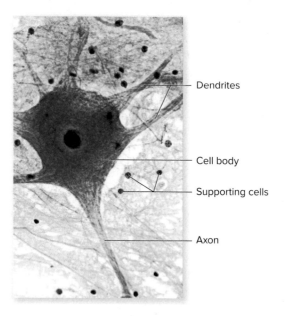

FIGURE 1.11   A photomicrograph of nerve tissue. A single neuron and numerous smaller supporting cells can be seen. **AP|R**
© Ed Reschke

the nucleus and serves as the metabolic center of the cell. The dendrites (literally, "branches") are highly branched cytoplasmic extensions of the cell body that receive input from other neurons or from receptor cells. The axon is a single cytoplasmic extension of the cell body that can be quite long (up to a few feet in length). It is specialized for conducting nerve impulses from the cell body to another neuron or to an effector (muscle or gland) cell.

Neuroglia do not conduct impulses but instead serve to bind neurons together, modify the extracellular environment of the nervous system, and influence the nourishment and electrical activity of neurons. In recent years, neuroglia have been shown to cooperate with neurons in chemical neurotransmission (chapter 7), and to have many other roles in the normal physiology (as well as disease processes) of the brain and spinal cord. Neuroglia are about five times more abundant than neurons and can divide by mitosis throughout life, whereas mature neurons do not divide.

Neurons and neuroglia are discussed in detail in chapter 7.

Epithelial Tissue

Epithelial tissue consists of cells that form **membranes,** which cover and line the body surfaces, and of **glands,** which are derived from these membranes. There are two categories of glands. *Exocrine glands* (from the Greek *exo* = outside)

secrete chemicals through a duct that leads to the outside of a membrane, and thus to the outside of a body surface. *Endocrine glands* (from the Greek *endon* = within) secrete chemicals called *hormones* into the blood. Endocrine glands are discussed in chapter 11.

Epithelial Membranes

Epithelial membranes are classified according to the number of their layers and the shape of the cells in the upper layer (table 1.3). Epithelial cells that are flattened in shape are **squamous;** those that are as wide as they are tall are **cuboidal;** and those that are taller than they are wide are **columnar** (fig. 1.12*a–c*). Those epithelial membranes that are only one cell layer thick are known as **simple epithelia;** those that are composed of a number of layers are **stratified epithelia.**

Epithelial membranes cover all body surfaces and line the cavity (lumen) of every hollow organ. Thus, epithelial membranes provide a barrier between the external environment and the internal environment of the body. Stratified epithelia are specialized to provide protection. Simple epithelia, in contrast, provide little protection; instead, they are specialized for transport of substances between the internal and external environments. In order for a substance to get into the body, it must pass through an epithelial membrane, and simple epithelia are specialized for this function. For example, a simple squamous

TABLE 1.3 | Summary of Epithelial Membranes

Type	Structure and Function	Location
Simple Epithelia	Single layer of cells; function varies with type	Covering visceral organs; linings of body cavities, tubes, and ducts
Simple squamous epithelium	Single layer of flattened, tightly bound cells; diffusion and filtration	Capillary walls; pulmonary alveoli of lungs; covering visceral organs; linings of body cavities
Simple cuboidal epithelium	Single layer of cube-shaped cells; excretion, secretion, or absorption	Surface of ovaries; linings of kidney tubules, salivary ducts, and pancreatic ducts
Simple columnar epithelium	Single layer of nonciliated, tall, column-shaped cells; protection, secretion, and absorption	Lining of most of digestive tract
Simple ciliated columnar epithelium	Single layer of ciliated, column-shaped cells; transportive role through ciliary motion	Lining of uterine tubes
Pseudostratified ciliated columnar epithelium	Single layer of ciliated, irregularly shaped cells; many goblet cells; protection, secretion, ciliary movement	Lining of respiratory passageways
Stratified Epithelia	Two or more layers of cells; function varies with type	Epidermis of skin; linings of body openings, ducts, and urinary bladder
Stratified squamous epithelium (keratinized)	Numerous layers containing keratin, with outer layers flattened and dead; protection	Epidermis of skin
Stratified squamous epithelium (nonkeratinized)	Numerous layers lacking keratin, with outer layers moistened and alive; protection and pliability	Linings of oral and nasal cavities, vagina, and anal canal
Stratified cuboidal epithelium	Usually two layers of cube-shaped cells; strengthening of luminal walls	Large ducts of sweat glands, salivary glands, and pancreas
Transitional epithelium	Numerous layers of rounded, nonkeratinized cells; distension	Walls of ureters, part of urethra, and urinary bladder

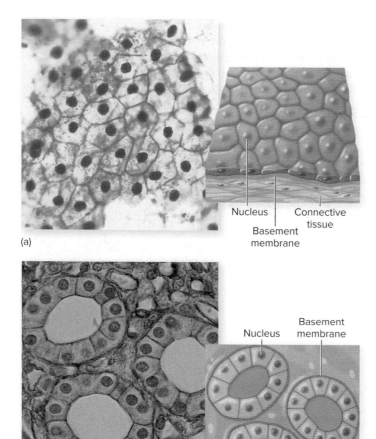

(a)

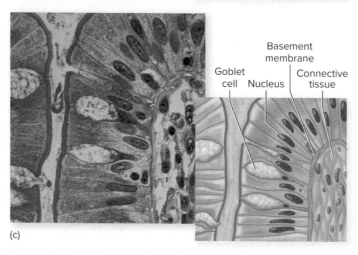

(b)

(c)

FIGURE 1.12 **Different types of simple epithelia.**
(a) Simple squamous, (b) simple cuboidal, and (c) simple columnar epithelia. The basement membrane, composed of proteins and cabohydrates, binds the epithelial membrane to the underlying connective tissue. (a) © Ray Simons/Science Source; (b) © Ray Simons/Science Source; (c) © Ed Reschke

epithelium in the lungs allows the rapid passage of oxygen and carbon dioxide between the air (external environment) and blood (internal environment). A simple columnar epithelium in the small intestine, as another example, allows digestion

products to pass from the intestinal lumen (external environment) to the blood (internal environment).

Dispersed among the columnar epithelial cells are specialized unicellular glands called *goblet cells* that secrete mucus. The columnar epithelial cells in the uterine tubes of females and in the respiratory passages contain numerous *cilia* (hairlike structures, described in chapter 3) that can move in a coordinated fashion and aid the functions of these organs.

The epithelial lining of the esophagus and vagina that provides protection for these organs is a stratified squamous epithelium (fig. 1.13). This is a *nonkeratinized* membrane, and all layers consist of living cells. The *epidermis* of the skin, by contrast, is *keratinized,* or *cornified* (fig. 1.14). Because the epidermis is dry and exposed to the potentially desiccating effects of the air, the surface is covered with dead cells that are filled with a water-resistant protein known as *keratin.* This protective layer is constantly flaked off from the surface of the skin and therefore must be constantly replaced by the division of cells in the deeper layers of the epidermis.

The constant loss and renewal of cells is characteristic of epithelial membranes. The entire epidermis is completely replaced every two weeks; the stomach lining is renewed

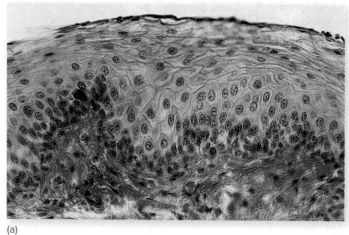

(a)

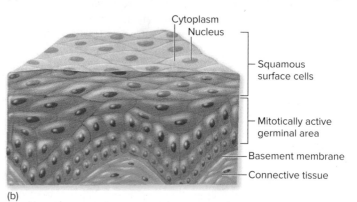

(b)

FIGURE 1.13 **A stratified squamous nonkeratinized epithelial membrane.** This is a photomicrograph (a) and illustration (b) of the epithelial lining of the vagina. AP|R
(a) © Victor P. Eroschenko

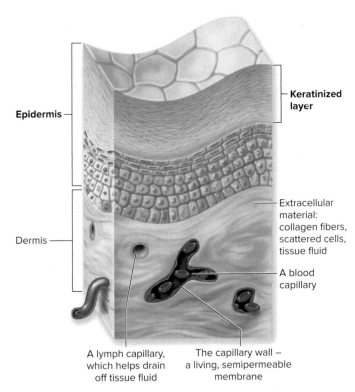

Epidermis

Keratinized layer

Dermis

Extracellular material: collagen fibers, scattered cells, tissue fluid

A blood capillary

A lymph capillary, which helps drain off tissue fluid

The capillary wall – a living, semipermeable membrane

FIGURE 1.14   The epidermis is a stratified, squamous, keratinized epithelium. The upper cell layers are dead and impregnated with the protein keratin, producing a cornified epithelial membrane, which is supported by layers of living cells. The epidermis is nourished by blood vessels located in the loose connective tissue of the dermis. AP|R

every two to three days. Examination of the cells that are lost, or "exfoliated," from the outer layer of epithelium lining the female reproductive tract is a common procedure in gynecology (as in the Pap smear).

In order to form a strong membrane that is effective as a barrier at the body surfaces, epithelial cells are very closely packed and are joined together by *intercellular junctions* collectively called **junctional complexes** (chapter 6; see fig. 6.22). There is no room for blood vessels between adjacent epithelial cells. The epithelium must therefore receive nourishment from the tissue beneath, which has large intercellular spaces that can accommodate blood vessels and nerves. This underlying tissue is called *connective tissue*. Epithelial membranes are attached to the underlying connective tissue by a layer of proteins and polysaccharides known as the **basement membrane.** This layer can be observed only under the microscope using specialized staining techniques.

Basement membranes are believed to induce a polarity to the cells of epithelial membranes; that is, the top (apical) portion of epithelial cells has different structural and functional components than the bottom (basal) portion. This is important in many physiological processes. For example, substances are transported in specific directions across simple epithelia (discussed in chapter 6; see fig. 6.21). In stratified epithelia, only the basal (bottom) layer of cells is on the basement membrane,

and it is these cells that undergo mitosis to form new epithelial cells to replace those lost from the top. Scientists recently demonstrated that when these basal cells divide, one of the daughter cells is attached to the basement membrane (renewing the basal cell population), while the other is not. The daughter cell that is "unstuck" from the basement membrane differentiates (becomes specialized) and migrates upward in the stratified epithelium.

CLINICAL APPLICATION

Basement membranes consist primarily of the structural protein known as *collagen* (see fig. 1.17). The type of collagen in basement membranes is a large protein assembled from six different subunits. **Alport syndrome** is a genetic disorder of the collagen subunits that, among other problems, results in damage to the glomeruli (the filtering units) of the kidneys. This is one of the most common causes of kidney failure. In **Goodpasture's syndrome,** the collagen in the basement membranes of the glomeruli and the lungs is attacked by the person's own antibodies, leading to both kidney and lung disease.

Exfoliative cytology is the collection and examination of epithelial cells that are shed and collected by mechanical scraping of the membranes, washing of the membranes, or aspiration of body fluids containing the shed cells. Microscopic examination of these *desquamated* (shed) cells, for example in a **Pap smear,** may reveal a malignancy.

Exocrine Glands

Exocrine glands are derived from cells of epithelial membranes. The secretions of these cells are passed to the outside of the epithelial membranes (and hence to the surface of the body) through *ducts.* This is in contrast to *endocrine glands,* which lack ducts and which therefore secrete into capillaries within the body (fig. 1.15). The structure of endocrine glands will be described in chapter 11.

The secretory units of exocrine glands may be simple tubes, or they may be modified to form clusters of units around branched ducts (fig. 1.16). These clusters, or **acini,** are often surrounded by tentacle-like extensions of *myoepithelial cells* that contract and squeeze the secretions through the ducts. The rate of secretion and the action of myoepithelial cells are subject to neural and endocrine regulation.

Examples of exocrine glands in the skin include the lacrimal (tear) glands, sebaceous glands (which secrete oily sebum into hair follicles), and sweat glands. There are two types of sweat glands. The more numerous, the *eccrine* (or *merocrine*) *sweat glands,* secrete a dilute salt solution that serves in thermoregulation (evaporation cools the skin). The *apocrine sweat glands,* located in the axillae (underarms) and pubic region, secrete a protein-rich fluid and become more active during puberty. This provides nourishment for bacteria that produce the characteristic odor of this type of sweat.

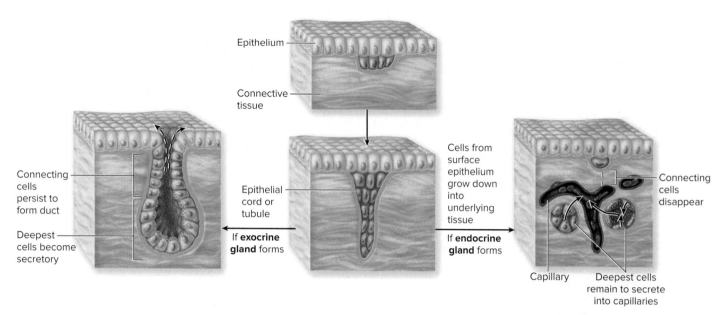

FIGURE 1.15 **The formation of exocrine and endocrine glands from epithelial membranes.** Note that exocrine glands retain a duct that can carry their secretion to the surface of the epithelial membrane, whereas endocrine glands are ductless.

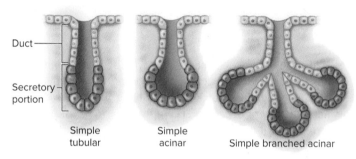

FIGURE 1.16 **The structure of exocrine glands.** Exocrine glands may be simple invaginations of epithelial membranes, or they may be more complex derivatives. **AP|R**

All of the glands that secrete into the digestive tract are also exocrine. This is because the lumen of the digestive tract is a part of the external environment, and secretions of these glands go to the outside of the membrane that lines this tract. Mucous glands are located throughout the length of the digestive tract. Other relatively simple glands of the tract include salivary glands, gastric glands, and simple tubular glands in the intestine.

The *liver* and *pancreas* are exocrine (as well as endocrine) glands, derived embryologically from the digestive tract. The exocrine secretion of the pancreas—pancreatic juice—contains digestive enzymes and bicarbonate and is secreted into the small intestine via the pancreatic duct. The liver produces and secretes bile (an emulsifier of fat) into the small intestine via the gallbladder and bile duct.

Exocrine glands are also prominent in the reproductive system. The female reproductive tract contains numerous mucus-secreting exocrine glands. The male accessory sex organs—the *prostate* and *seminal vesicles*—are exocrine glands that contribute to semen. The testes and ovaries (the gonads) are both endocrine and exocrine glands. They are endocrine because they secrete sex steroid hormones into the blood; they are exocrine because they release gametes (ova and sperm) into the reproductive tracts.

Connective Tissue

Connective tissue is characterized by large amounts of extracellular material between the different types of connective tissue cells. The extracellular material, called the *extracellular matrix,* varies in the four primary types of connective tissues: (1) connective tissue proper; (2) cartilage; (3) bone; and (4) blood. **Blood** is classified as a type of connective tissue because about half its volume is an extracellular fluid, the blood plasma (chapter 13, section 13.1).

Connective tissue proper, in which the extracellular matrix consists of protein fibers and a proteinaceous, gel-like *ground substance,* is divided into subtypes. In *loose connective tissue,* protein fibers composed of *collagen* are scattered loosely in the ground substance (fig. 1.17), which provides space for the presence of blood vessels, nerve fibers, and other structures (see the dermis of the skin, shown in fig. 1.14, as an example). *Dense regular connective tissues* are those in which collagen fibers are oriented parallel to each other and densely packed in the extracellular matrix, leaving little room for cells and ground substance (fig. 1.18). Examples of dense regular connective tissues include tendons (connecting bone to bone) and ligaments (connecting bones together at joints). *Dense irregular connective tissues,* forming tough capsules and sheaths around organs, contain densely packed collagen bundles arranged in various orientations that resist forces applied from different directions.

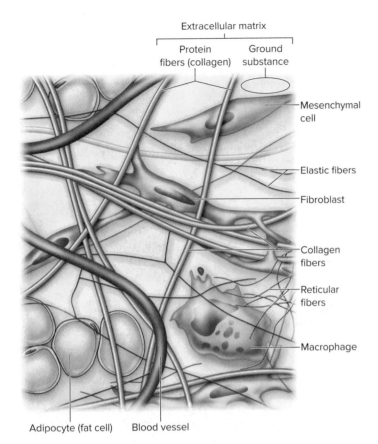

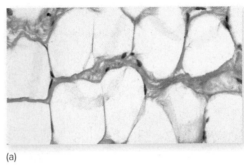

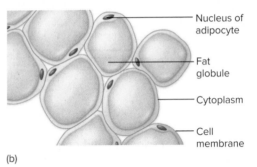

(a)

(b)

FIGURE 1.19   **Adipose tissue.** Each adipocyte contains a large, central globule of fat surrounded by the cytoplasm of the adipocyte. (*a*) Photomicrograph and (*b*) illustration of adipose tissue. AP|R (a) © McGraw-Hill Education/Al Telser, photographer

FIGURE 1.17   **Loose connective tissue.** This illustration shows the cells and protein fibers characteristic of connective tissue proper. The ground substance is the extracellular background material, against which the different protein fibers can be seen. The macrophage is a phagocytic connective tissue cell, which can be derived from monocytes (a type of white blood cell). AP|R

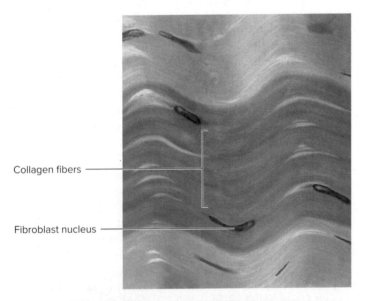

FIGURE 1.18   **Dense regular connective tissue.** In this photomicrograph, the collagen fibers in a tendon are packaged densely into parallel groups. The ground substance is in the tiny spaces between the collagen fibers. AP|R © McGraw-Hill Education

Adipose tissue is a specialized type of loose connective tissue. In each adipose cell, or *adipocyte,* the cytoplasm is stretched around a central globule of fat (fig. 1.19). The synthesis and breakdown of fat are accomplished by enzymes within the cytoplasm of the adipocytes.

Cartilage consists of cells, called *chondrocytes,* surrounded by a semisolid ground substance that imparts elastic properties to the tissue. Cartilage is a type of supportive and protective tissue commonly called "gristle." It forms the precursor to many bones that develop in the fetus and persists at the articular (joint) surfaces on the bones at all movable joints in adults.

Bone is produced as concentric layers, or *lamellae,* of calcified material laid around blood vessels. The bone-forming cells, or *osteoblasts,* surrounded by their calcified products, become trapped within cavities called *lacunae.* The trapped cells, which are now called *osteocytes,* remain alive because they are nourished by "lifelines" of cytoplasm that extend from the cells to the blood vessels in *canaliculi* (little canals). The blood vessels lie within central canals, surrounded by concentric rings of bone lamellae with their trapped osteocytes. These units of bone structure are called *osteons* (fig. 1.20).

The *dentin* of a tooth (fig. 1.21) is similar in composition to bone, but the cells that form this calcified tissue are located in the pulp (composed of loose connective tissue). These cells send cytoplasmic extensions, called *dentinal tubules,* into the dentin. Dentin, like bone, is thus a living tissue that can be remodeled in response to stresses. The cells that form the outer *enamel* of a tooth, by contrast, are lost as the tooth erupts. Enamel is a

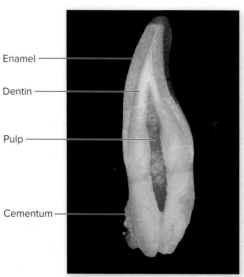

Enamel ——

Dentin ——

Pulp ——

Cementum ——

FIGURE 1.21   A cross section of a tooth showing pulp, dentin, and enamel. The root of the tooth is covered by cementum, a calcified connective tissue that helps to anchor the tooth in its bony socket. © Southern Illinois University/Science Source

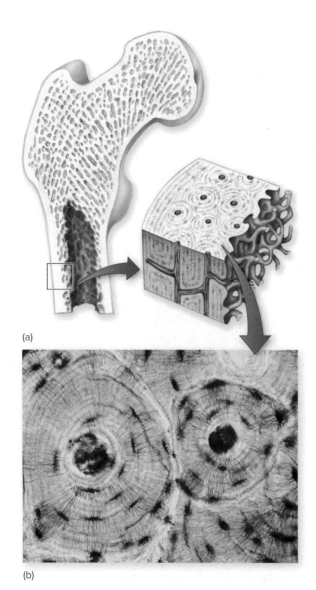

(a)

(b)

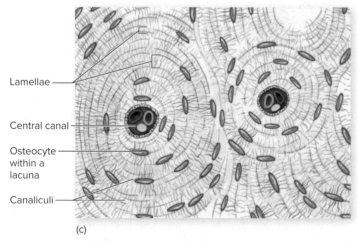

Lamellae ——

Central canal ——

Osteocyte within a lacuna ——

Canaliculi ——

(c)

FIGURE 1.20   The structure of bone. (*a*) A diagram of a long bone, (*b*) a photomicrograph showing osteons, and (*c*) a diagram of osteons. Within each central canal, an artery (red), a vein (blue), and a nerve (yellow) is illustrated. **AP|R** (b) © Ed Reschke

highly calcified material, harder than bone or dentin, that cannot be regenerated; artificial "fillings" are therefore required to patch holes in the enamel.

CHECKPOINTS

6a. List the four primary tissues and describe the distinguishing features of each type.

6b. Compare and contrast the three types of muscle tissue.

6c. Describe the different types of epithelial membranes and state their locations in the body.

7a. Explain why exocrine and endocrine glands are considered epithelial tissues and distinguish between these two types of glands.

7b. Describe the different types of connective tissues and explain how they differ from one another in their content of extracellular material.

1.4 ORGANS AND SYSTEMS

Organs are composed of two or more primary tissues that serve the different functions of the organ. The skin is an organ that has numerous functions provided by its constituent tissues.

LEARNING OUTCOMES

After studying this section, you should be able to:

8. Use the skin as an example to describe how the different primary tissues compose organs.

9. Identify the body fluid compartments.

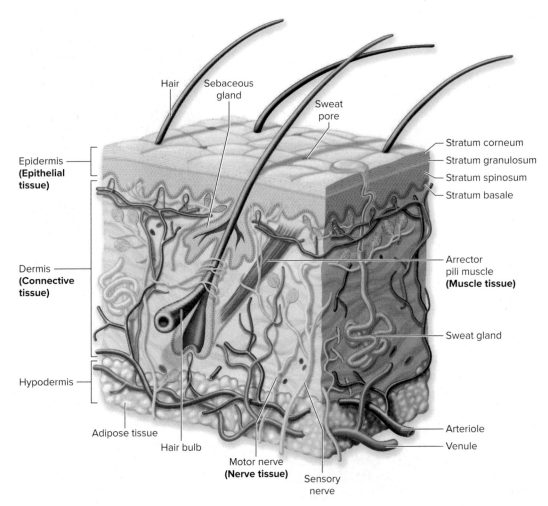

Hair   Sebaceous
        gland

Sweat
pore

Stratum corneum

Stratum granulosum

Stratum spinosum

Stratum basale

Epidermis
**(Epithelial
tissue)**

Dermis
**(Connective
tissue)**

Arrector
pili muscle
(Muscle tissue)

Sweat gland

Hypodermis

Adipose tissue

Hair bulb

Motor nerve
(Nerve tissue)

Sensory
nerve

Arteriole

Venule

FIGURE 1.22   A diagram of the skin. The skin is an organ that contains all four types of primary tissues. **AP|R**

An **organ** is a structure composed of at least two, and usually all four, primary tissues. The largest organ in the body, in terms of surface area, is the skin (fig. 1.22). In this section, the numerous functions of the skin serve to illustrate how primary tissues cooperate in the service of organ physiology.

An Example of an Organ: The Skin

The cornified *epidermis* protects the skin against water loss and against invasion by disease-causing organisms. Invaginations of the epithelium into the underlying connective tissue *dermis* create the exocrine glands of the skin. These include hair follicles (which produce the hair), sweat glands, and sebaceous glands. The secretion of sweat glands cools the body by evaporation and produces odors that, at least in lower animals, serve as sexual attractants. Sebaceous glands secrete oily sebum into hair follicles, which transport the sebum to the surface of the skin. Sebum lubricates the cornified surface of the skin, helping to prevent it from drying and cracking.

The skin is nourished by blood vessels within the dermis. In addition to blood vessels, the dermis contains wandering

white blood cells and other types of cells that protect against invading disease-causing organisms. It also contains nerve fibers and adipose (fat) cells; however, most of the adipose cells are grouped together to form the *hypodermis* (a layer beneath the dermis). Although adipose cells are a type of connective tissue, masses of fat deposits throughout the body are referred to as *adipose tissue.*

Sensory nerve endings within the dermis mediate the cutaneous sensations of touch, pressure, heat, cold, and pain. Motor nerve fibers in the skin stimulate effector organs, resulting in, for example, the secretions of exocrine glands and contractions of the arrector pili muscles, which attach to hair follicles and surrounding connective tissue (producing goose bumps). The degree of constriction or dilation of cutaneous blood vessels— and therefore the rate of blood flow—is also regulated by motor nerve fibers.

The epidermis itself is a dynamic structure that can respond to environmental stimuli. The rate of its cell division—and consequently the thickness of the cornified layer—increases under the stimulus of constant abrasion. This produces calluses. The skin also protects itself against the dangers of ultraviolet light by increasing its production of *melanin* pigment, which absorbs ultraviolet light while producing a tan. In addition, the skin is an endocrine gland; it synthesizes and secretes vitamin D (derived from cholesterol under the influence of ultraviolet light), which functions as a hormone.

The architecture of most organs is similar to that of the skin. Most are covered by an epithelium that lies immediately over a connective tissue layer. The connective tissue contains blood vessels, nerve endings, scattered cells for fighting infection, and possibly glandular tissue as well. If the organ is hollow—as with the digestive tract or blood vessels—the lumen is also lined with an epithelium overlying a connective tissue layer. The presence, type, and distribution of muscle tissue and nerve tissue vary in different organs.

Stem Cells

The different tissues of an organ are composed of cells that are highly specialized, or *differentiated.* The process of differentiation begins during embryonic development, when the fertilized ovum, or *zygote,* divides to produce three embryonic tissue layers, or *germ layers: ectoderm, mesoderm,* and *endoderm* (chapter 20; see fig. 20.45*a*). During the course of embryonic and fetal development, the three germ layers give rise to the four primary tissues and their subtypes.

The zygote is *totipotent*—it can produce all of the different specialized cells of the body. After a number of cell divisions, when the embryo is at the stage where it implants into the mother's uterus, its cells are *pluripotent*—they can produce all of the body cells except those that contribute to the placenta, and so have been called **embryonic stem cells.** As development proceeds through successive cell divisions, the cells become increasingly differentiated and lose the ability to form unrelated cell types. However, genes are not lost during differentiation. This was first demonstrated in the 1960s by a British scientist, who found that he could produce a fully formed frog by transplanting the nucleus from a differentiated frog intestinal cell into the cytoplasm of an enucleated ovum. In 2006 and later, Japanese scientists found that they could transform differentiated fibroblasts into a pluripotent state, similar to embryonic stem cells, by treating the fibroblasts with a few specific regulatory molecules. The 2012 Nobel Prize in Physiology or Medicine was awarded for these discoveries.

Because the specialized cells have a limited life span, many organs retain small populations of cells that are less differentiated and more able to divide to become the specialized cell types within the organ. These less-differentiated cells are known as **adult stem cells.** In the red bone marrow, for example, the stem cell population gives rise to all of the different blood cells—red blood cells, white blood cells, and platelets (chapter 13). Similarly, there are stem cells in the brain (chapter 8), skeletal muscles (chapter 12), and intestine (chapter 18).

Scientists have recently discovered that there are also stem cells in the bulge region of the hair follicle (fig. 1.23). These stem cells form keratinocytes, which migrate down to the lower germinal matrix of the hair follicle and divide to form the hair

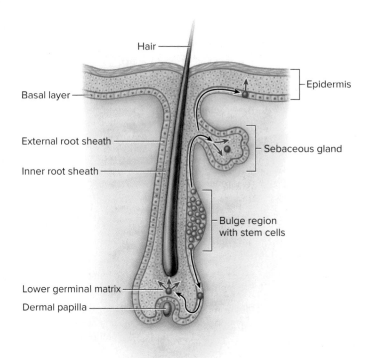

FIGURE 1.23   The bulge region of the hair follicle with stem cells. Stem cells in this region migrate to form the differentiated cells of the hair follicle, sebaceous gland, and epidermis. AP|R

shaft and root sheath. Other stem cells in the region of the hair follicle just above the bulge form new sebaceous gland cells, which have a high turnover. Skin wounds stimulate the migration of stem cells from the hair follicles into the skin between follicles to promote healing of the wounded skin.

The bulge region also contains melanocyte stem cells, which migrate to the lower germinal matrix of the follicle and give the hair its color. Scientists have now shown that graying of the hair with age is caused by loss of the melanocyte stem cells in the bulge of the hair follicles. The melanocyte stem cells appeared to be present in most of the hair follicles of people aged 20 to 30 and absent from most hair follicles of people aged 70 and older.

As demonstrated by the stem cells in the bulge of the hair follicle, adult stem cells can form a variety of related cell types; the adult stem cells are therefore described as *multipotent.* The topics of embryonic and adult stem cells are discussed in more detail in the context of embryonic development (chapter 20, section 20.6).

Systems

Organs that are located in different regions of the body and that perform related functions are grouped into **systems.** These include the integumentary system, nervous system, endocrine system, skeletal system, muscular system, circulatory system, immune system, respiratory system, urinary system, alimentary (digestive) system, and reproductive system (table 1.4). By means of numerous regulatory mechanisms, these systems work together to maintain the life and health of the entire organism.

TABLE 1.4 | Organ Systems of the Body

System	Major Organs	Primary Functions
Integumentary	Skin, hair, nails	Protection, thermoregulation
Nervous	Brain, spinal cord, nerves	Regulation of other body systems
Endocrine	Hormone-secreting glands, such as the pituitary, thyroid, and adrenal glands	Secretion of regulatory molecules called hormones
Skeletal	Bones, cartilages	Movement and support
Muscular	Skeletal muscles	Movements of the skeleton
Circulatory	Heart, blood vessels, lymphatic vessels	Movement of blood and lymph
Immune	Red bone marrow, lymphoid organs	Defense of the body against invading pathogens
Respiratory	Lungs, airways	Gas exchange
Urinary	Kidneys, ureters, urethra	Regulation of blood volume and composition
Alimentary	Mouth, stomach, intestine, liver, gallbladder, pancreas	Breakdown of food into molecules that enter the body
Reproductive	Gonads, external genitalia, associated glands and ducts	Continuation of the human species

Body-Fluid Compartments

Tissues, organs, and systems can all be divided into two major parts, or compartments. The **intracellular compartment** is that part inside the cells; the **extracellular compartment** is that part outside the cells. Both compartments consist primarily of water—they are said to be *aqueous.* About 65% of the total body water is in the intracellular compartment, while about 35% is in the extracellular compartment. The two compartments are separated by the plasma membrane surrounding each cell (chapter 3, section 3.1).

The extracellular compartment is subdivided into two parts. One part is the *blood plasma,* the fluid portion of the blood. The other is the fluid that bathes the cells within the organs of the body. This is called *interstitial fluid.* In most parts of the body, blood plasma and interstitial fluid communicate freely through blood capillaries. The kidneys regulate the volume and composition of the blood plasma, and thus, indirectly, the fluid volume and composition of the entire extracellular compartment.

There is also selective communication between the intracellular and extracellular compartments through the movement of molecules and ions through the cell membrane, as described in chapter 6. This is how cells obtain the molecules they need for life and how they eliminate waste products.

CHECKPOINTS

8a. State the location of each type of primary tissue in the skin.

8b. Describe the functions of nervous, muscle, and connective tissue in the skin.

8c. Describe the functions of the epidermis and explain why this tissue is called "dynamic."

9. Distinguish between the intracellular and extracellular compartments and explain their significance.

CLINICAL INVESTIGATION   SUMMARY

Homeostasis of body temperature and blood glucose is maintained by physiological mechanisms that operate in negative feedback loops. Health requires that the body maintain homeostasis, and this can be assessed clinically by specific measurements. If Linda's fasting blood glucose were above the normal range, it could indicate a disease called diabetes mellitus. In this disease, the hormone insulin does not lower the blood glucose concentration adequately in response to a rise in blood glucose.

You can solve another medical mystery now by going to the Connect site for this text. Look for Feedback Control of Blood Glucose, which is based on this chapter's material. The more you play with physiological concepts, the better you will understand them. The Clinical Investigation boxes at the end of subsequent chapters will provide additional Clinical Investigations on the Connect site that relate to those chapters.

SUMMARY

1.1 Introduction to Physiology   2

A. Physiology is the study of how cells, tissues, and organs function.

1. In the study of physiology, cause-and-effect sequences are emphasized.

2. Knowledge of physiological mechanisms is deduced from data obtained experimentally.

B. The science of physiology shares knowledge with the related sciences of pathophysiology and comparative physiology.

1. Pathophysiology is concerned with the functions of diseased or injured body systems and is based on knowledge of how normal systems function, which is the focus of physiology.

2. Comparative physiology is concerned with the physiology of animals other than humans and shares much information with human physiology.

C. All of the information in this book has been gained by applications of the scientific method. This method has three essential characteristics:

1. It is assumed that the subject under study can ultimately be explained in terms we can understand.

2. Descriptions and explanations are honestly based on observations of the natural world and can be changed as warranted by new observations.

3. Humility is an important characteristic of the scientific method; the scientist must be willing to change his or her theories when warranted by the weight of the evidence.

1.2 Homeostasis and Feedback Control   4

A. Homeostasis refers to the dynamic constancy of the internal environment.

1. Homeostasis is maintained by mechanisms that act through negative feedback loops.

   a. A negative feedback loop requires (1) a sensor that can detect a change in the internal environment and (2) an effector that can be activated by the sensor.

   b. In a negative feedback loop, the effector acts to cause changes in the internal environment that compensate for the initial deviations that were detected by the sensor.

2. Positive feedback loops serve to amplify changes and may be part of the action of an overall negative feedback mechanism.

3. The nervous and endocrine systems provide extrinsic regulation of other body systems and act to maintain homeostasis.

4. The secretion of hormones is stimulated by specific chemicals and is inhibited by negative feedback mechanisms.

B. Effectors act antagonistically to defend the set point against deviations in any direction.

1.3 The Primary Tissues   10

A. The body is composed of four types of primary tissues: muscle, nerve, epithelial, and connective tissues.

1. There are three types of muscle tissue: skeletal, cardiac, and smooth muscle.

   a. Skeletal and cardiac muscle are striated muscle tissue.

   b. Smooth muscle is found in the walls of the internal organs.

2. Nerve tissue is composed of neurons and neuroglia.

   a. Neurons are specialized for the generation and conduction of electrical impulses.

   b. Neuroglia provide the neurons with anatomical and functional support.

3. Epithelial tissue includes membranes and glands.

   a. Epithelial membranes cover and line the body surfaces, and their cells are tightly joined by intercellular junctions.

   b. Epithelial membranes may be simple or stratified, and their cells may be squamous, cuboidal, or columnar.

   c. Exocrine glands, which secrete into ducts, and endocrine glands, which lack ducts and secrete hormones into the blood, are derived from epithelial membranes.

4. Connective tissue is characterized by large intercellular spaces that contain extracellular material.

   a. Connective tissue proper is categorized into subtypes, including loose, dense fibrous, adipose, and others.

   b. Cartilage, bone, and blood are classified as connective tissues because their cells are widely spaced with abundant extracellular material between them.

1.4 Organs and Systems   17

A. Organs are units of structure and function that are composed of at least two, and usually all four, of the primary types of tissues.

1. The skin is a good example of an organ.

   a. The epidermis is a stratified squamous keratinized epithelium that protects underlying structures and produces vitamin D.

   b. The dermis is an example of loose connective tissue.

   c. Hair follicles, sweat glands, and sebaceous glands are exocrine glands located within the dermis.

   d. Sensory and motor nerve fibers enter the spaces within the dermis to innervate sensory organs and smooth muscles.

   e. The arrector pili muscles that attach to the hair follicles are composed of smooth muscle.

2. Organs that are located in different regions of the body and that perform related functions are grouped into systems. These include, among others, the circulatory system, alimentary system, and endocrine system.

3. Many organs contain adult stem cells, which are able to differentiate into a number of related cell types.

   a. Because of their limited flexibility, adult stem cells are described as multipotent, rather than as totipotent or pluripotent.

   b. For example, the bulge region of a hair follicle contains stem cells that can become keratinocytes, epithelial cells, and melanocytes; the loss of the melanocyte stem cells causes graying of the hair.

B. The fluids of the body are divided into two major compartments.

1. The intracellular compartment refers to the fluid within cells.

2. The extracellular compartment refers to the fluid outside of cells; extracellular fluid is subdivided into plasma (the fluid portion of the blood) and interstitial fluid.

REVIEW ACTIVITIES

Test your knowledge

1. Glands are derived from
   - **a.** nerve tissue.
   - **b.** connective tissue.
   - **c.** muscle tissue.
   - **d.** epithelial tissue.

2. Cells joined tightly together are characteristic of
   - **a.** nerve tissue.
   - **b.** connective tissue.
   - **c.** muscle tissue.
   - **d.** epithelial tissue.

3. Cells are separated by large extracellular spaces in
   - **a.** nerve tissue.
   - **b.** connective tissue.
   - **c.** muscle tissue.
   - **d.** epithelial tissue.

4. Blood vessels and nerves are usually located within
   - **a.** nerve tissue.
   - **b.** connective tissue.
   - **c.** muscle tissue.
   - **d.** epithelial tissue.

5. Most organs are composed of
   - **a.** epithelial tissue.
   - **b.** muscle tissue.
   - **c.** connective tissue.
   - **d.** all of these.

6. Sweat is secreted by exocrine glands. This means that
   - **a.** it is produced by endocrine cells.
   - **b.** it is a hormone.
   - **c.** it is secreted into a duct.
   - **d.** it is produced outside the body.

7. Which of these statements about homeostasis is *true?*
   - **a.** The internal environment is maintained absolutely constant.
   - **b.** Negative feedback mechanisms act to correct deviations from a normal range within the internal environment.
   - **c.** Homeostasis is maintained by turning effectors on and off.
   - **d.** All of these are true.

8. In a negative feedback loop, the effector produces changes that are
   - **a.** in the same direction as the change produced by the initial stimulus.
   - **b.** opposite in direction to the change produced by the initial stimulus.
   - **c.** unrelated to the initial stimulus.

9. A hormone called parathyroid hormone acts to help raise the blood calcium concentration. According to the principles of negative feedback, an effective stimulus for parathyroid hormone secretion would be
   - **a.** a fall in blood calcium.
   - **b.** a rise in blood calcium.

10. Which of these consists of dense parallel arrangements of collagen fibers?
    - **a.** skeletal muscle tissue
    - **b.** nerve tissue
    - **c.** tendons
    - **d.** dermis of the skin

11. The act of breathing raises the blood oxygen level, lowers the blood carbon dioxide concentration, and raises the blood pH. According to the principles of negative feedback, sensors that regulate breathing should respond to
    - **a.** a rise in blood oxygen.
    - **b.** a rise in blood pH.
    - **c.** a rise in blood carbon dioxide concentration.
    - **d.** all of these.

12. Adult stem cells, such as those in the bone marrow, brain, or hair follicles, can best be described as _____, whereas embryonic stem cells are described as

    _____.
    - **a.** totipotent; pluripotent
    - **b.** pluripotent; multipotent
    - **c.** multipotent; pluripotent
    - **d.** totipotent; multipotent

Test Your Understanding

13. Describe the structure of the various epithelial membranes and explain how their structures relate to their functions.

14. Compare bone, blood, and the dermis of the skin in terms of their similarities. What are the major structural differences between these tissues?

15. Describe the role of antagonistic negative feedback processes in the maintenance of homeostasis.

16. Using insulin as an example, explain how the secretion of a hormone is controlled by the effects of that hormone's actions.

17. Describe the steps in the development of pharmaceutical drugs and evaluate the role of animal research in this process.

18. Why is Claude Bernard considered the father of modern physiology? Why is the concept he introduced so important in physiology and medicine?

Test Your Analytical Ability

19. What do you think would happen if most of your physiological regulatory mechanisms were to operate by positive feedback rather than by negative feedback? Would life even be possible?

20. Examine figure 1.5 and determine when the compensatory physiological responses began to act, and how many minutes they required to restore the initial set point of blood glucose concentration. Comment on the importance of quantitative measurements in physiology.

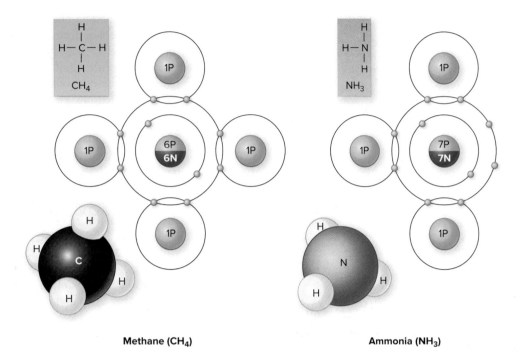

Methane (CH₄)　　　　　　**Ammonia (NH₃)**

FIGURE 2.3 **The molecules methane and ammonia represented in three different ways.** Notice that a bond between 2 atoms consists of a pair of shared electrons (the electrons from the outer shell of each atom).

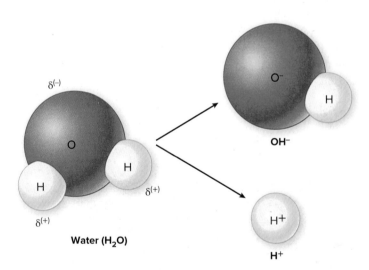

Water (H₂O)

FIGURE 2.4 **A model of a water molecule showing its polar nature.** Notice that the oxygen side of the molecule is relatively more negative, whereas the hydrogen side is relatively more positive. Polar covalent bonds are weaker than nonpolar covalent bonds. As a result, some water molecules ionize to form a hydroxide ion (OH⁻) and a hydrogen ion (H⁺).

the electrons are not shared at all. The first atom loses electrons so that its number of electrons becomes smaller than its number of protons; it becomes positively charged. Atoms or molecules that have positive or negative charges are called **ions.** Positively charged ions are called *cations* because they move toward the negative pole, or cathode, in an electric field.

The second atom now has more electrons than it has protons and becomes a negatively charged ion, or *anion* (so called because it moves toward the positive pole, or anode, in an electric field). The cation and anion then attract each other to form an **ionic compound.**

Common table salt, sodium chloride (NaCl), is an example of an ionic compound. Sodium, with a total of 11 electrons, has 2 in its first shell, 8 in its second shell, and only 1 in its third shell. Chlorine, conversely, is 1 electron short of completing its outer shell of 8 electrons. The lone electron in sodium's outer shell is attracted to chlorine's outer shell. This creates a chloride ion (represented as Cl⁻) and a sodium ion (Na⁺). Although table salt is shown as NaCl, it is actually composed of Na⁺Cl⁻ (fig. 2.5).

Ionic compounds are held together by the attraction of opposite charges, and these compounds easily dissociate (separate) when they are dissolved in water. Dissociation of NaCl, for example, yields Na⁺ and Cl⁻. Each of these ions attracts polar water molecules; the negative ends of water molecules are attracted to the Na⁺, and the positive ends of water molecules are attracted to the Cl⁻ (fig. 2.6). The water molecules that surround these ions, in turn, attract other molecules of water to form *hydration spheres* around each ion.

It is the formation of hydration spheres that makes an ion or a molecule soluble in water. Glucose, amino acids, and many other organic molecules are water soluble because hydration spheres can form around atoms of oxygen, nitrogen, and phosphorus, which are joined by polar covalent bonds to other atoms in the molecule. Such molecules are said to be **hydrophilic.** By contrast, molecules composed primarily of nonpolar covalent

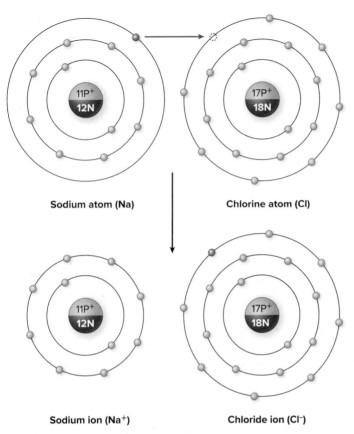

Sodium atom (Na)   **Chlorine atom (Cl)**

Sodium ion (Na⁺)   **Chloride ion (Cl⁻)**

FIGURE 2.5   The reaction of sodium with chlorine to produce sodium and chloride ions. The positive sodium and negative chloride ions attract each other, producing the ionic compound sodium chloride (NaCl).

bonds, such as the hydrocarbon chains of fat molecules, have few charges and thus cannot form hydration spheres. They are insoluble in water and appear repelled by water molecules (because the water molecules preferentially bond with each other; fig. 2.7). For this reason, nonpolar molecules are said to be **hydrophobic** ("water fearing").

Hydrogen Bonds

When a hydrogen atom forms a polar covalent bond with an atom of oxygen or nitrogen, the hydrogen gains a slight positive charge as its electron is pulled toward the other atom. This other atom is thus described as being *electronegative*. Because the hydrogen has a slight positive charge, it will have a weak attraction for a second electronegative atom (oxygen or nitrogen) that may be located near it. This weak attraction is called a **hydrogen bond.** Hydrogen bonds are usually shown with dashed or dotted lines (fig. 2.7) to distinguish them from strong covalent bonds, which are shown with solid lines.

Although each hydrogen bond is relatively weak, the sum of their attractive forces is largely responsible for the folding and bending of long organic molecules such as proteins and for the holding together of the two strands of a DNA molecule (described in section 2.4). Hydrogen bonds can also be formed between adjacent water molecules (fig. 2.7). The hydrogen bonding between water molecules is responsible for many of the biologically important properties of water, including its *surface tension* and its ability to be pulled as a column through narrow channels in a process called *capillary action*.

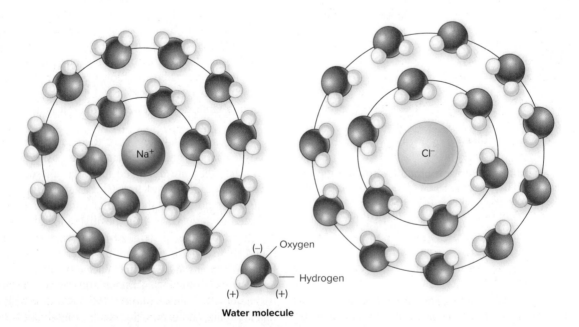

(−) ── Oxygen

── Hydrogen

(+)   (+)

Water molecule

FIGURE 2.6   How NaCl dissolves in water. The negatively charged oxygen-ends of water molecules are attracted to the positively charged Na⁺, whereas the positively charged hydrogen-ends of water molecules are attracted to the negatively charged Cl⁻. Other water molecules are attracted to this first concentric layer of water, forming hydration spheres around the sodium and chloride ions. **AP|R**

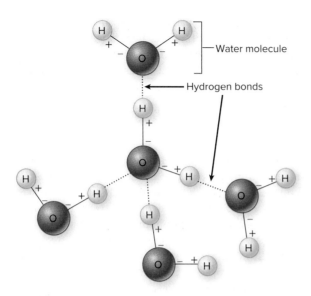

FIGURE 2.7   Hydrogen bonds between water molecules. The oxygen atoms of water molecules are weakly joined together by the attraction of the negatively charged oxygen for the positively charged hydrogen. These weak bonds are called hydrogen bonds. **AP|R**

Acids, Bases, and the pH Scale

The bonds in water molecules joining hydrogen and oxygen atoms together are, as previously discussed, polar covalent bonds. Although these bonds are strong, a small proportion of them break as the electron from the hydrogen atom is completely transferred to oxygen. When this occurs, the water molecule ionizes to form a *hydroxide ion* (OH⁻) and a hydrogen ion (H⁺), which is simply a free proton (see fig. 2.4). A proton released in this way does not remain free for long, however, because it is attracted to the electrons of oxygen atoms in water molecules. This forms a *hydronium ion,* shown by the formula H_3O^+. For the sake of clarity in the following discussion, however, H^+ will be used to represent the ion resulting from the ionization of water.

Ionization of water molecules produces equal amounts of OH⁻ and H⁺. Only a small proportion of water molecules ionize, so the concentrations of H⁺ and OH⁻ are each equal to only 10^{-7} molar (M). The term *molar* is a unit of concentration, described in chapter 6; for hydrogen, 1 molar equals 1 gram per liter. A solution with 10^{-7} molar hydrogen ions, produced by the ionization of water molecules in which the H⁺ and OH⁻ concentrations are equal, is said to be **neutral.**

A solution that has a higher H⁺ concentration than that of water is called *acidic;* one with a lower H⁺ concentration is called *basic,* or *alkaline.* An **acid** is defined as a molecule that can release protons (H⁺) into a solution; it is a "proton donor." A **base** can be a molecule such as ammonia (NH_3) that can combine with H⁺ (to form NH_4^+, ammonium ion). More commonly, it is a molecule such as NaOH that can ionize to produce

TABLE 2.2 | Common Acids and Bases

Acid	Symbol	Base	Symbol
Hydrochloric acid	HCl	Sodium hydroxide	NaOH
Phosphoric acid	H_3PO_4	Potassium hydroxide	KOH
Nitric acid	HNO_3	Calcium hydroxide	$Ca(OH)_2$
Sulfuric acid	H_2SO_4	Ammonium hydroxide	NH_4OH
Carbonic acid	H_2CO_3		

a negatively charged ion (hydroxide, OH⁻), which, in turn, can combine with H⁺ (to form H_2O, water). A base thus removes H⁺ from solution; it is a "proton acceptor," thereby lowering the H⁺ concentration of the solution. Examples of common acids and bases are shown in table 2.2.

pH

The H⁺ concentration of a solution is usually indicated in pH units on a pH scale that runs from 0 to 14. The pH value is equal to the logarithm of 1 over the H⁺ concentration:

$$pH = \log \frac{1}{[H^+]}$$

where [H⁺] = molar H⁺ concentration. This can also be expressed as $pH = -\log[H^+]$.

Pure water has a H⁺ concentration of 10^{-7} molar at 25° C, and thus has a pH of 7 (neutral). Because of the logarithmic relationship, a solution with 10 times the hydrogen ion concentration (10^{-6} M) has a pH of 6, whereas a solution with one-tenth the H⁺ concentration (10^{-8} M) has a pH of 8. The pH value is easier to write than the molar H⁺ concentration, but it is admittedly confusing because it is *inversely related* to the H⁺ concentration. That is, a solution with a higher H⁺ concentration has a lower pH value, and one with a lower H⁺ concentration has a higher pH value. A strong acid with a high H⁺ concentration of 10^{-2} molar, for example, has a pH of 2, whereas a solution with only 10^{-10} molar H⁺ has a pH of 10. **Acidic solutions,** therefore, have a pH of less than 7 (that of pure water), whereas **basic (alkaline) solutions** have a pH between 7 and 14 (table 2.3).

Buffers

A **buffer** is a system of molecules and ions that acts to prevent changes in H⁺ concentration and thus serves to stabilize the pH of a solution. In blood plasma, for example, the pH is stabilized by the following reversible reaction involving the bicarbonate ion (HCO_3^-) and carbonic acid (H_2CO_3):

$$HCO_3^- + H^+ \rightleftarrows H_2CO_3$$

TABLE 2.3 | The pH Scale

	H^+ Concentration (Molar)*	pH	OH^- Concentration (Molar)*
Acids	1.0	0	10^{-14}
	0.1	1	10^{-13}
	0.01	2	10^{-12}
	0.001	3	10^{-11}
	0.0001	4	10^{-10}
	10^{-5}	5	10^{-9}
	10^{-6}	6	10^{-8}
Neutral	10^{-7}	7	10^{-7}
Bases	10^{-8}	8	10^{-6}
	10^{-9}	9	10^{-5}
	10^{-10}	10	0.0001
	10^{-11}	11	0.001
	10^{-12}	12	0.01
	10^{-13}	13	0.1
	10^{-14}	14	1.0

*Molar concentration is the number of moles of a solute dissolved in one liter. One mole is the atomic or molecular weight of the solute in grams. Since hydrogen has an atomic weight of one, 1 molar hydrogen is 1 gram of hydrogen per liter of solution.

The double arrows indicate that the reaction could go either to the right or to the left; the net direction depends on the concentration of molecules and ions on each side. If an acid (such as lactic acid) should release H^+ into the solution, for example, the increased concentration of H^+ would drive the equilibrium to the right and the following reaction would be promoted:

$$HCO_3^- + H^+ \rightarrow H_2CO_3$$

Notice that in this reaction, H^+ is taken out of solution. Thus, the H^+ concentration is prevented from rising (and the pH prevented from falling) by the action of bicarbonate buffer.

Blood pH

Lactic acid and other organic acids are produced by the cells of the body and secreted into the blood. Despite the release of H^+ by these acids, the arterial blood pH normally does not decrease but remains remarkably constant at pH 7.40 ± 0.05. This constancy is achieved, in part, by the buffering action of bicarbonate shown in the preceding equation. Bicarbonate serves as the major buffer of the blood.

Certain conditions could cause an opposite change in pH. For example, excessive vomiting that results in loss of gastric acid could cause the concentration of free H^+ in the blood to fall and the blood pH to rise. In this case, the reaction previously described could be reversed:

$$H_2CO_3 \rightarrow H^+ + HCO_3^-$$

The dissociation of carbonic acid yields free H^+, which helps to prevent an increase in pH. Bicarbonate ions and carbonic acid thus act as a *buffer pair* to prevent either decreases or increases in pH, respectively. This buffering action normally maintains the blood pH within the narrow range of 7.35 to 7.45.

If the arterial blood pH falls below 7.35, the condition is called *acidosis*. A blood pH of 7.20, for example, represents significant acidosis. Notice that acidotic blood need not be acidic (have a pH less than 7.00). An increase in blood pH above 7.45, conversely, is known as *alkalosis*. Acidosis and alkalosis are normally prevented by the action of the bicarbonate/carbonic acid buffer pair and by the functions of the lungs and kidneys. Regulation of blood pH is discussed in more detail in chapters 16 and 17.

Organic Molecules

Organic molecules are those molecules that contain the atoms carbon and hydrogen (in addition to other atoms). Because the carbon atom has 4 electrons in its outer shell, it must share 4 additional electrons by covalently bonding with other atoms to fill its outer shell with 8 electrons. The unique bonding requirements of carbon enable it to join with other carbon atoms to form chains and rings while still allowing the carbon atoms to bond with hydrogen and other atoms.

Most organic molecules in the body contain hydrocarbon chains and rings, as well as other atoms bonded to carbon. Two adjacent carbon atoms in a chain or ring may share one or two pairs of electrons. If the 2 carbon atoms share one pair of electrons, they are said to have a *single covalent bond;* this leaves each carbon atom free to bond with as many as 3 other atoms. If the 2 carbon atoms share two pairs of electrons, they have a *double covalent bond,* and each carbon atom can bond with a maximum of only 2 additional atoms (fig. 2.8).

The ends of some hydrocarbons are joined together to form rings. In the shorthand structural formulas for these molecules, the carbon atoms are not shown but are understood to be located at the corners of the ring. Some of these cyclic molecules have a double bond between 2 adjacent carbon atoms. Benzene and related molecules are shown as a six-sided ring with alternating double bonds. Such compounds are called **aromatic.** Because all of the carbons in an aromatic ring are equivalent, double bonds can be shown between any 2 adjacent carbons in the ring (fig. 2.9), or even as a circle within the hexagonal structure of carbons.

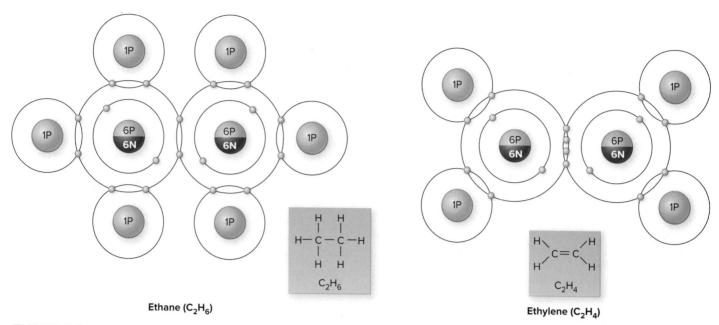

FIGURE 2.8   Single and double covalent bonds. Two carbon atoms may be joined by a single covalent bond (*left*) or a double covalent bond (*right*). In both cases, each carbon atom shares four pairs of electrons (has four bonds) to complete the 8 electrons required to fill its outer shell. **AP|R**

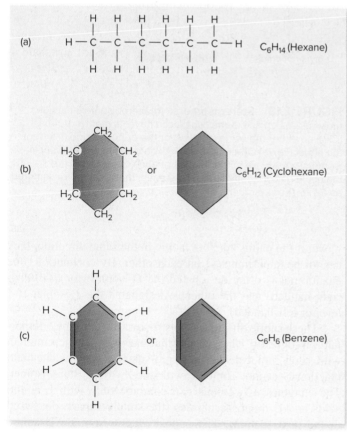

FIGURE 2.9   Different shapes of hydrocarbon molecules. Hydrocarbon molecules can be (*a*) linear or (*b*) cyclic or have (*c*) aromatic rings. **AP|R**

The hydrocarbon chain or ring of many organic molecules provides a relatively inactive molecular "backbone" to which more reactive groups of atoms are attached. Known as *functional groups* of the molecule, these reactive groups usually contain atoms of oxygen, nitrogen, phosphorus, or sulfur. They are largely responsible for the unique chemical properties of the molecule (fig. 2.10).

Classes of organic molecules can be named according to their functional groups. **Ketones,** for example, have a *carbonyl group* within the carbon chain. An organic molecule is an **alcohol** if it has a *hydroxyl group* bound to a hydrocarbon chain. All **organic acids** (acetic acid, citric acids, lactic acid, and others) have a *carboxyl group* (fig. 2.11).

A carboxyl group can be abbreviated COOH. This group is an acid because it can donate its proton (H^+) to the solution. Ionization of the OH part of COOH forms COO^- and H^+ (fig. 2.12). The ionized organic acid is designated with the suffix -*ate*. For example, when the carboxyl group of lactic acid ionizes, the molecule is called *lactate*. Both ionized and unionized forms of the molecule may exist together in a solution, with the proportion of each dependent on the pH of the solution. In the blood, the predominant form of this molecule is lactate.

Stereoisomers

Stereoisomers are molecules that have the same atoms in the same sequence, but differ from each other in the way their atoms are arranged three-dimensionally in space. Stereoisomers include (1) those that are designated as *cis* (Latin for

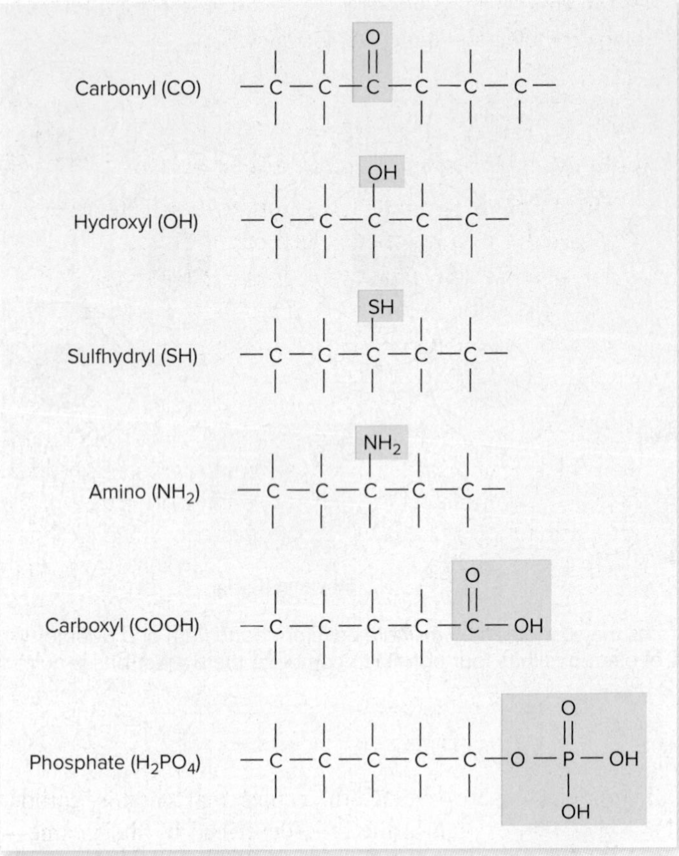

FIGURE 2.10 **Various functional groups of organic molecules.** The general symbol for any functional group is *R*. Specific functional groups are indicated within the gold rectangles.

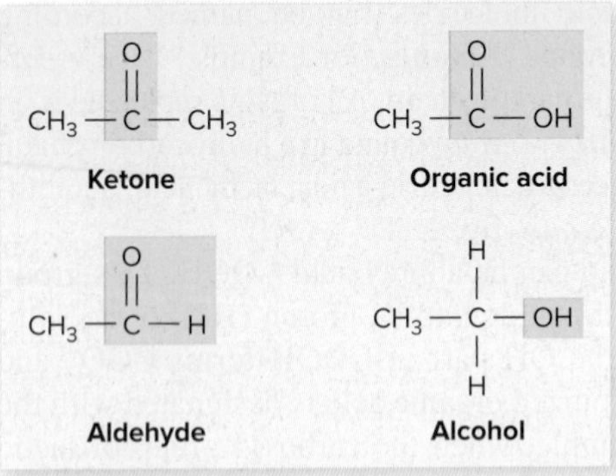

FIGURE 2.11 **Categories of organic molecules based on functional groups.** Acids, alcohols, and other types of organic molecules are characterized by specific functional groups.

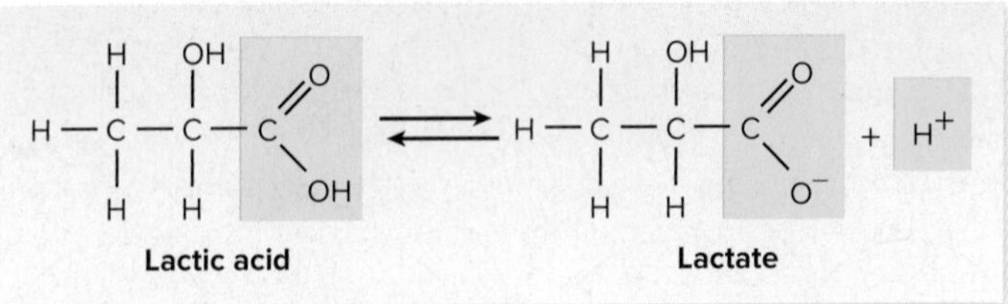

FIGURE 2.12 **The carboxyl group of an organic acid.** This group can ionize to yield a free proton, which is a hydrogen ion (H$^+$). This process is shown for lactic acid, with the double arrows indicating that the reaction is reversible.

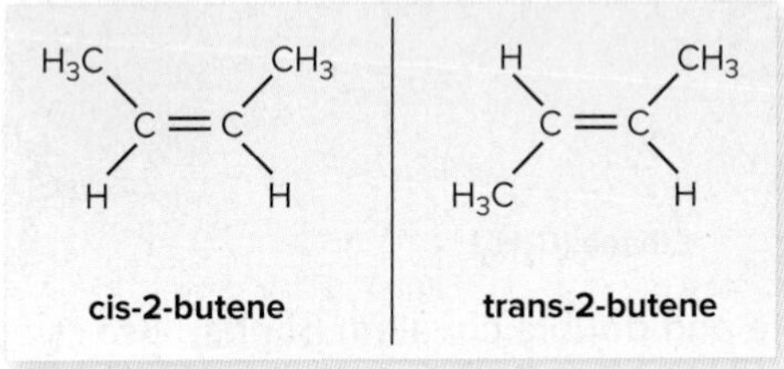

FIGURE 2.13 **Stereoisomers.** (*a*) Stereoisomers include molecules such as butene, which can have its methyl groups (CH$_3$) on either the same side (*cis*) or the opposite side (*trans*) of the molecule. (*b*) Other stereoisomers are enantiomers (optical isomers), which are mirror images of each other. Using the molecule glyceraldehyde as a reference, these can be designated as either *D* or *L* isomers.

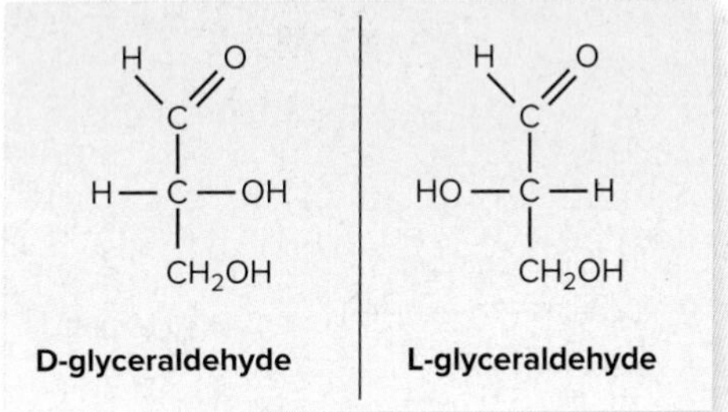

"on this side") and *trans* (meaning "across"), where two functional groups are located either on the same side (*cis*) or across from each other (*trans*) in the molecule (fig. 2.13*a*); and (2) **enantiomers,** which are mirror images of each other (fig. 2.13*b*). Enantiomers are like a left- and a right-hand glove; if the palms are both facing in the same direction, they cannot be superimposed on each other. By convention, one enantiomer is often designated the *D*-isomer (for *dextro,* or right-handed) and the other is designated the *L*-isomer (for *levo,* or left-handed).

These subtle differences in structure are extremely important biologically. They ensure that enzymes—which interact with such molecules in a stereo-specific way in chemical reactions—cannot combine with the "wrong" stereoisomer. The enzymes of all cells can combine only with L-amino acids and D-monosaccharides (the simple sugars, discussed in the next section). The opposite stereoisomers (D-amino acids and L-monosaccharides) cannot be used by any enzyme in metabolism.

In the mid-twentieth century the drug **thalidomide** was widely used by pregnant women to treat morning sickness, until scientists realized that it caused birth defects; it was banned in 1961. These **teratogenic effects** (those that cause abnormal embryonic development) were produced by only one of its enantiomers, but because thalidomide can convert to either of its enantiomers in the body, no form of thalidomide is safe for pregnant women. However, thalidomide and its derivatives, *lenalidomide* and *pomalidomide*, are now used as beneficial drugs for nonpregnant people in the treatment of *multiple myeloma* (a cancer of the bone marrow), *leprosy*, and *AIDS*.

Brian was concerned that the interconversion of thalidomide enantiomers (to form a *racemic mixture* of the two) in his body would have negative effects.

- Which enantiomers of amino acids and monosaccharides can be used by cells, and which cannot?
- Does the ability of thalidomide to convert to either of its enantiomers, and its teratogenic effects, pose a danger to Brian?

CHECKPOINTS

1. List the components of an atom and explain how they are organized. Explain why different atoms are able to form characteristic numbers of chemical bonds.

2. Describe the nature of nonpolar and polar covalent bonds, ionic bonds, and hydrogen bonds. Why are ions and polar molecules soluble in water?

3a. Define the terms *acidic, basic, acid,* and *base.* Also define *pH* and describe the relationship between pH and the H^+ concentration of a solution.

3b. Using chemical equations, explain how bicarbonate ion and carbonic acid function as a buffer pair.

4. Explain how carbon atoms can bond with each other and with atoms of hydrogen, oxygen, and nitrogen.

2.2 CARBOHYDRATES AND LIPIDS

Carbohydrates are a class of organic molecules that includes monosaccharides, disaccharides, and polysaccharides. All of these molecules are based on a characteristic ratio of carbon, hydrogen, and oxygen atoms. Lipids constitute a category of diverse organic molecules that share the physical property of being nonpolar, and thus insoluble in water.

LEARNING OUTCOMES

After studying this section, you should be able to:

5. Identify the different types of carbohydrates and lipids, and give examples of each type.

6. Explain how dehydration synthesis and hydrolysis reactions occur in carbohydrates and triglycerides.

7. Describe the nature of phospholipids and prostaglandins.

Carbohydrates and lipids are similar in many ways. Both groups of molecules consist primarily of the atoms carbon, hydrogen, and oxygen, and both serve as major sources of energy in the body (accounting for most of the calories consumed in food). Carbohydrates and lipids differ, however, in some important aspects of their chemical structures and physical properties. Such differences significantly affect the functions of these molecules in the body.

Carbohydrates

Carbohydrates are organic molecules that contain carbon, hydrogen, and oxygen in the ratio described by their name—*carbo* (carbon) and *hydrate* (water, H_2O). The general formula for monosaccharides (the building blocks of carbohydrates) is $C_nH_{2n}O_n$. This formula indicates that there are twice as many hydrogens as carbon or oxygen atoms, with the number of each indicated by the subscript *n*.

Monosaccharides, Disaccharides, and Polysaccharides

Carbohydrates include simple sugars, or **monosaccharides,** and longer molecules that contain a number of monosaccharides joined together. The suffix *-ose* denotes a sugar molecule; the term *hexose,* for example, refers to a six-carbon monosaccharide with the formula $C_6H_{12}O_6$. This formula is adequate for some purposes, but it does not distinguish between related hexose sugars, which are *structural isomers* of each other. The structural isomers glucose, galactose, and fructose, for example, are monosaccharides that have the same ratio of atoms arranged in slightly different ways (fig. 2.14).

Two monosaccharides can be joined covalently to form a **disaccharide,** or double sugar. Common disaccharides include table sugar, or *sucrose* (composed of glucose and fructose); milk sugar, or *lactose* (composed of glucose and galactose); and malt sugar, or *maltose* (composed of two glucose molecules). When numerous monosaccharides are joined together, the resulting molecule is called a **polysaccharide.**

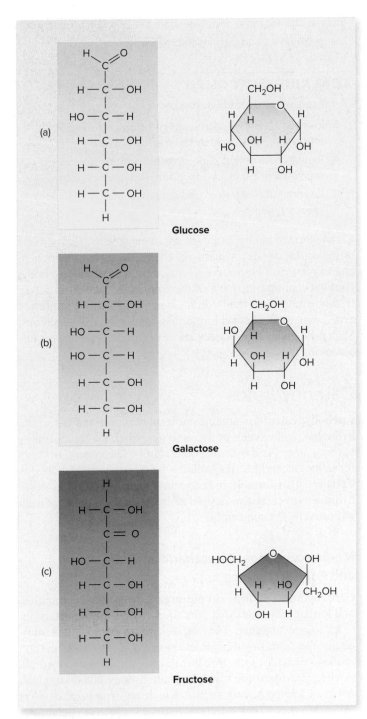

FIGURE 2.14 Structural formulas for three hexose sugars. These are (a) glucose, (b) galactose, and (c) fructose. All three have the same ratio of atoms—$C_6H_{12}O_6$. The representations on the left more clearly show the atoms in each molecule, while the ring structures on the right more accurately reflect the way these atoms are arranged.

See the *Test Your Quantitative Ability* section of the *Review Activities* at the end of this chapter.

The major polysaccharides are chains of repeating glucose subunits. *Starch* is a plant product formed by the bonding together of thousands of glucose subunits into long chains, and **glycogen** (sometimes called animal starch) is similar, but more highly branched (fig. 2.15). Animals have the enzymes to digest the bonds (chemically called alpha-1,4 glycosidic bonds) between adjacent glucose subunits of these polysaccharides. *Cellulose* (produced by plants) is also a polysaccharide of glucose, but the bonds joining its glucose subunits are oriented differently (forming beta-1,4 glycosidic bonds) than those in starch or glycogen. Because of this, our digestive enzymes cannot hydrolyze cellulose into its glucose subunits. However, animals such as cows, horses, and sheep—which eat grasses—can digest cellulose because they have symbiotic bacteria with the necessary enzymes in their digestive tracts. *Chitin* (poly-N-acetylglucosamine) is a polysaccharide similar to cellulose (with beta-1,4 glycosidic bonds) but with amine-containing groups in the glucose subunits. Chitin forms the exoskeleton of arthropods such as insects and crustaceans.

Many cells store carbohydrates for use as an energy source, as described in chapter 5. If many thousands of separate monosaccharide molecules were stored in a cell, however, their high concentration would draw an excessive amount of water into the cell, damaging or even killing it. The net movement of water through membranes is called osmosis, and is discussed in chapter 6. Cells that store carbohydrates for energy minimize this osmotic damage by instead joining the glucose molecules together to form the polysaccharides starch or glycogen. Because there are fewer of these larger molecules, less water is drawn into the cell by osmosis (see chapter 6).

Dehydration Synthesis and Hydrolysis

In the formation of disaccharides and polysaccharides, the separate subunits (monosaccharides) are bonded together covalently by a type of reaction called **dehydration synthesis,** or **condensation.** In this reaction, which requires the participation of specific enzymes (chapter 4), a hydrogen atom is removed from one monosaccharide and a hydroxyl group (OH) is removed from another. As a covalent bond is formed between the two monosaccharides, water (H_2O) is produced. Dehydration synthesis reactions are illustrated in figure 2.16.

When a person eats disaccharides or polysaccharides, or when the stored glycogen in the liver and muscles is to be used by tissue cells, the covalent bonds that join monosaccharides to form disaccharides and polysaccharides must be broken. These *digestion reactions* occur by means of **hydrolysis.** Hydrolysis (from the Greek *hydro* = water; *lysis* = break) is the reverse of dehydration synthesis. When a covalent bond joining two monosaccharides is broken, a water molecule provides the atoms needed to complete their structure. The water molecule is split, and the resulting hydrogen atom is added to one of the

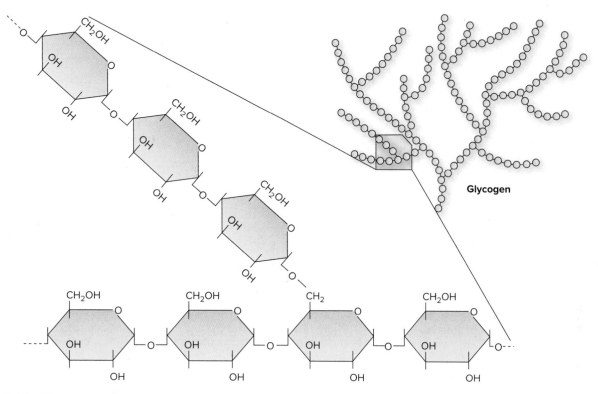

FIGURE 2.15   The structure of glycogen. Glycogen is a polysaccharide composed of glucose subunits joined together to form a large, highly branched molecule.

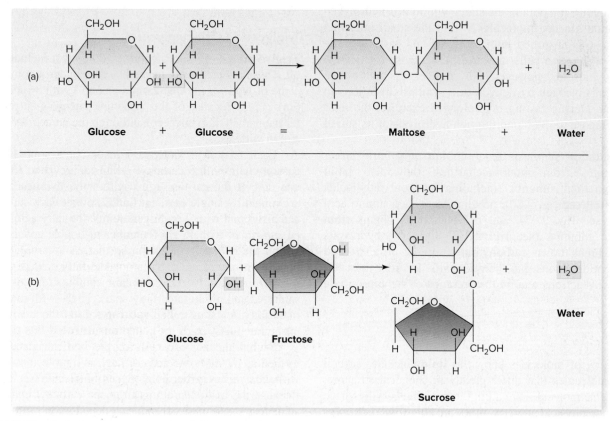

FIGURE 2.16   Dehydration synthesis of disaccharides. The two disaccharides formed here are (*a*) maltose and (*b*) sucrose (table sugar). Notice that a molecule of water is produced as the disaccharides are formed.

See *Test Your Quantitative Ability* section of the Review Activities at the end of this chapter.

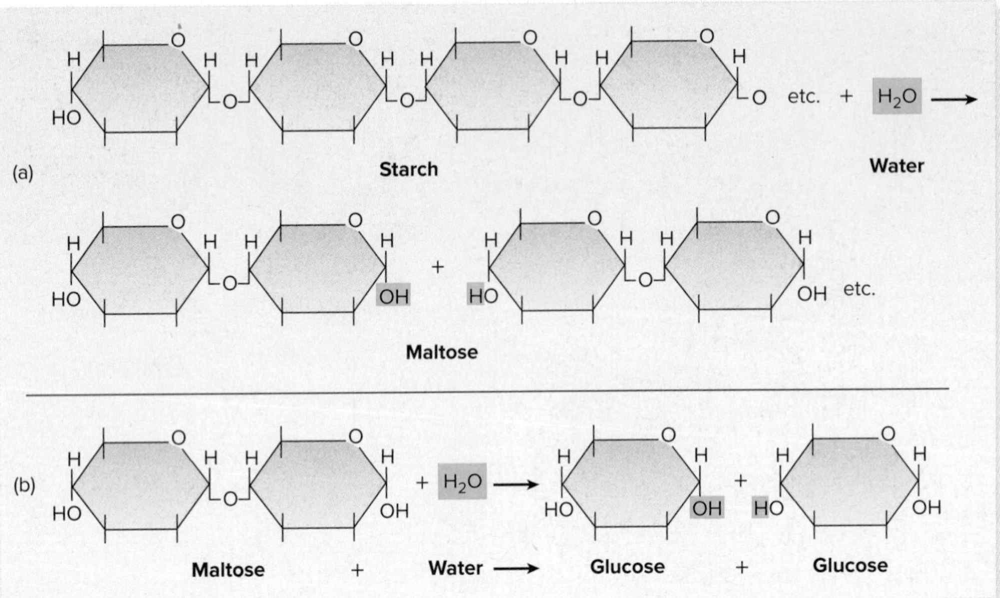

FIGURE 2.17 **The hydrolysis of starch.** The polysaccharide is first hydrolyzed into (*a*) disaccharides (maltose) and then into (*b*) monosaccharides (glucose). Notice that as the covalent bond between the subunits breaks, a molecule of water is split. In this way, the hydrogen atom and hydroxyl group from the water are added to the ends of the released subunits.

free glucose molecules as the hydroxyl group is added to the other (fig. 2.17).

When you eat a potato, the starch within it is hydrolyzed into separate glucose molecules within the small intestine. This glucose is absorbed into the blood and carried to the tissues. Some tissue cells may use this glucose for energy. Liver and muscles, however, can store excess glucose in the form of glycogen by dehydration synthesis reactions in these cells. During fasting or prolonged exercise, the liver can add glucose to the blood through hydrolysis of its stored glycogen.

Dehydration synthesis reactions not only build larger carbohydrates from monosaccharides, they also build lipids from their subunits (including fat from fatty acids and glycerol; see fig. 2.20), proteins from their amino acid subunits (see fig. 2.27), and polynucleotide chains from nucleotide subunits (see fig. 2.31). Similarly, hydrolysis reactions break down carbohydrates, lipids, proteins, and polynucleotide chains into their subunits. In order to occur, all of these reactions require the presence of the appropriate enzymes.

Lipids

The category of molecules known as **lipids** includes several types of molecules that differ greatly in chemical structure. These diverse molecules are all in the lipid category by virtue of a common physical property—they are all *insoluble in polar solvents* such as water. This is because lipids consist primarily of hydrocarbon chains and rings, which are nonpolar and therefore hydrophobic. Although lipids are insoluble in water,

they can be dissolved in nonpolar solvents such as ether, benzene, and related compounds.

Triglyceride (Triacylglycerol)

Triglyceride is the subcategory of lipids that includes fat and oil. These molecules are formed by the condensation of 1 molecule of *glycerol* (a three-carbon alcohol) with 3 molecules of *fatty acids*. Because of this structure, chemists currently prefer the name **triacylglycerol,** although the name *triglyceride* is still in wide use.

Each fatty acid molecule consists of a nonpolar hydrocarbon chain with a carboxyl group (abbreviated COOH) on one end. If the carbon atoms within the hydrocarbon chain are joined by single covalent bonds so that each carbon atom can also bond with 2 hydrogen atoms, the fatty acid is said to be *saturated*. If there are a number of double covalent bonds within the hydrocarbon chain, so that each carbon atom can bond with only 1 hydrogen atom, the fatty acid is said to be *unsaturated*. Triglycerides contain combinations of different saturated and unsaturated fatty acids. Those with mostly saturated fatty acids are called **saturated fats;** those with mostly unsaturated fatty acids are called **unsaturated fats** (fig. 2.18).

Within the adipose cells of the body, triglycerides are formed as the carboxyl ends of fatty acid molecules condense with the hydroxyl groups of a glycerol molecule (see fig. 2.20). Because the hydrogen atoms from the carboxyl ends of fatty acids form water molecules during dehydration synthesis, fatty acids that are combined with glycerol can no longer release H^+ and function as acids. For this reason, triglycerides are described as *neutral fats*.

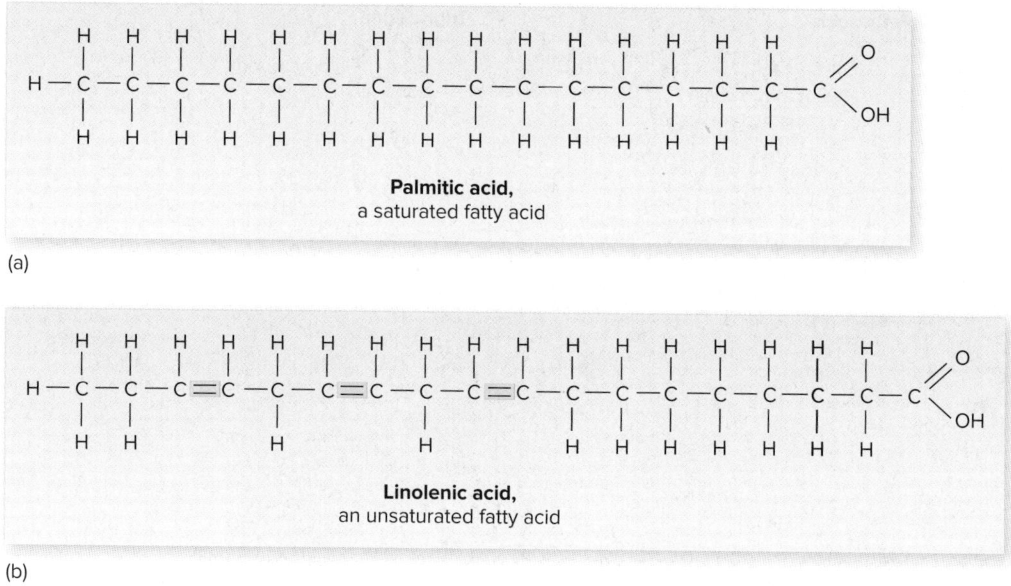

FIGURE 2.18 Structural formulas for fatty acids. (*a*) The formula for saturated fatty acids and (*b*) the formula for unsaturated fatty acids. Double bonds, which are points of unsaturation, are highlighted in yellow.

(a)

Palmitic acid,
a saturated fatty acid

(b)

Linolenic acid,
an unsaturated fatty acid

EXERCISE APPLICATION

Studies show that endurance exercise lowers plasma triglycerides and may also lower total cholesterol, effects that can be aided by choosing a proper diet. The American Heart Association recommends that people limit the amount of *saturated fat* they eat to less than 7%, and the amount of **trans fats** they eat to less than 1%, of the total number of calories they consume per day. This is because both saturated fats and trans fats may raise the total blood cholesterol levels, which contributes to **atherosclerosis** (hardening of the arteries). Trans fats not only raise the level of *LDL cholesterol* (the "bad cholesterol" that contributes to atherosclerosis), they also lower the amount of *HDL cholesterol* (the "good cholesterol" that helps protect against atherosclerosis). Trans fats additionally raise the plasma triglyceride levels and have other effects that promote the formation of plaques in the arteries that are characteristic of atherosclerosis. The Food and Drug Administration (FDA) now requires that all manufacturers list trans fats on their food labels.

The major food sources of saturated fats are meats and dairy; for example, beef fat is 52% saturated, and butter fat is 66% saturated. Trans fats are produced artificially by partially hydrogenating vegetable oils (this is how margarine is made), and they are common in fried and baked foods. In *trans fatty acids,* the single hydrogen atoms bonded to each carbon located on either side of the double bond are located on the opposite (*trans*) sides of the molecule (fig. 2.19, *right*), which causes the carbon atoms to form a straight chain. By contrast, in most naturally occurring fatty acids, the hydrogens are located on the same, or *cis*, side (fig. 2.19, *left*), causing the fatty acid chain to bend at the double bonds and produce a sawtooth pattern.

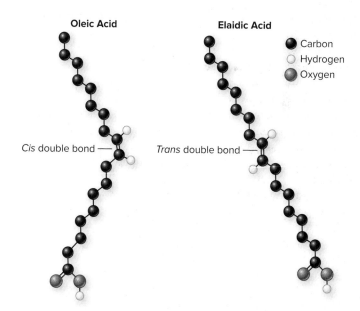

Oleic Acid   Elaidic Acid

● Carbon
○ Hydrogen
● Oxygen

Cis double bond ——   *Trans* double bond ——

FIGURE 2.19 The structure of cis and trans fatty acids. Oleic acid is a naturally occurring fatty acid with one double bond. Notice that both hydrogen atoms (yellow) on the carbons that share this double bond are on the same side of the molecule—this is called the *cis* configuration. The *cis* configuration makes this naturally occurring fatty acid bend. The fatty acid on the right is the same size and also has one double bond, but its hydrogens here are on opposite sides of the molecule, in the *trans* configuration. This makes the fatty acid stay straight, more like a saturated fatty acid. Note that only these hydrogens and the ones on the carboxyl groups (*bottom*) are shown. Those carbons that are joined by single bonds are also each bonded to 2 hydrogen atoms, but those hydrogens are not illustrated.

Fatty acids

Triglyceride

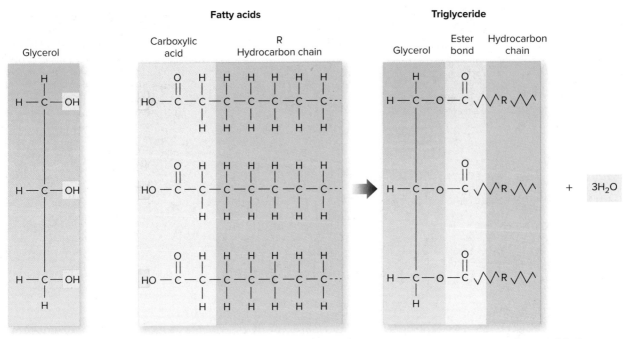

FIGURE 2.20   The formation of a triglyceride (triacylglycerol) molecule from glycerol and three fatty acids by dehydration synthesis reactions.   A molecule of water is produced as an ester bond forms between each fatty acid and the glycerol. Sawtooth lines represent hydrocarbon chains, generally 16 to 22 carbons long, which are symbolized by an *R*.

Ketone Bodies

Hydrolysis of triglycerides within adipose tissue releases *free fatty acids* into the blood. Free fatty acids can be used as an immediate source of energy by many organs; they can also be converted by the liver into derivatives called **ketone bodies** (fig. 2.21). These include four-carbon-long acidic molecules (acetoacetic acid and β-hydroxybutyric acid) and acetone (the solvent in nailpolish remover). A rapid breakdown of fat, as may occur during strict low-carbohydrate diets and in uncontrolled diabetes mellitus, results in elevated levels of ketone bodies in the blood. This condition,

called **ketosis,** is beneficial under these conditions because the brain can obtain energy from ketone bodies when adequate plasma glucose is unavailable. If there are sufficient amounts of ketone bodies in the blood to lower the blood pH, the condition is called **ketoacidosis.** Severe ketoacidosis, which may occur in diabetes mellitus, can lead to coma and death.

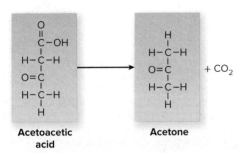

Acetoacetic acid

Acetone

$+ CO_2$

FIGURE 2.21   Ketone bodies. Acetoacetic acid, an acidic ketone body, can spontaneously decarboxylate (lose carbon dioxide) to form acetone. Acetone is a volatile ketone body that escapes in the exhaled breath, thereby lending a "fruity" smell (more the odor of nail polish remover) to the breath of people with ketosis (elevated blood ketone bodies).

CLINICAL INVESTIGATION   CLUES

When Brian's urine is tested in the laboratory, they discover that he has *ketonuria* (an elevated level of ketone bodies in his urine).

- What are ketone bodies, and how do they originate?
- What is the relationship between Brian's ketonuria and his weight loss?

Phospholipids

The group of lipids known as **phospholipids** includes a number of different categories of lipids, all of which contain a phosphate group. The most common type of phospholipid molecule is one in which the three-carbon alcohol molecule glycerol is attached to two fatty acid molecules. The third carbon atom of the glycerol is attached to a phosphate group, and the phosphate group, in turn, is bound to other molecules. If the phosphate group is attached to a nitrogen-containing choline

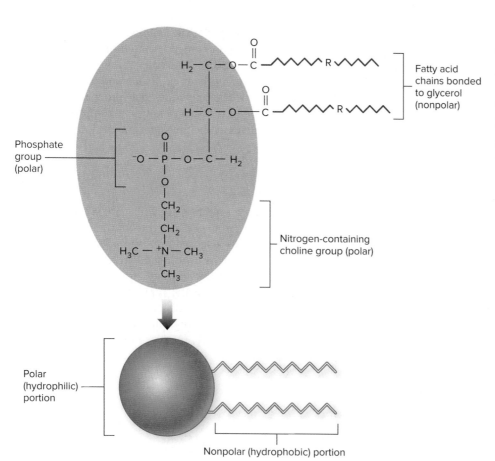

Phosphate group (polar)

Fatty acid chains bonded to glycerol (nonpolar)

Nitrogen-containing choline group (polar)

Polar (hydrophilic) portion

Nonpolar (hydrophobic) portion

FIGURE 2.22 **The structure of lecithin.** Lecithin is also called phosphatidylcholine, where choline is the nitrogen-containing portion of the molecule. (Interestingly, choline is also part of an important neurotransmitter known as acetylcholine, discussed in chapter 7.) The detailed structure of the phospholipid (*top*) is usually shown in simplified form (*bottom*), where the circle represents the polar portion and the sawtoothed lines the nonpolar portion of the molecule.

molecule, the phospholipid thus formed is known as **lecithin** (or *phosphatidylcholine*). The bottom part of figure 2.22 shows a simple way of illustrating the structure of a phospholipid—the parts of the molecule capable of ionizing (and thus becoming charged) are shown as a sphere, whereas the nonpolar parts of the molecule are represented by sawtooth lines. Molecules that are part polar and part nonpolar, such as phospholipids and bile acids (which are derived from cholesterol), are described as **amphipathic** molecules.

Phospholipids are the major component of cell membranes; their amphipathic nature allows them to form a double layer with their polar portions facing water on each side of the membrane (chapter 3). When phospholipids are mixed in water, they tend to group together so that their polar parts face the surrounding water molecules (fig. 2.23). Such aggregates of molecules are called **micelles.** Bile acids (which are not phospholipids, but are amphipathic molecules derived from cholesterol) form similar micelles in the small intestine (chapter 18, section 18.5). The amphipathic nature of phospholipids (part polar, part nonpolar) allows them to alter the interaction of water molecules and thus to decrease the surface tension of water. This function of phospholipids makes them **surfactants** (surface-active agents). The surfactant effect of phospholipids prevents the lungs from collapsing due to surface tension forces (chapter 16, section 16.2).

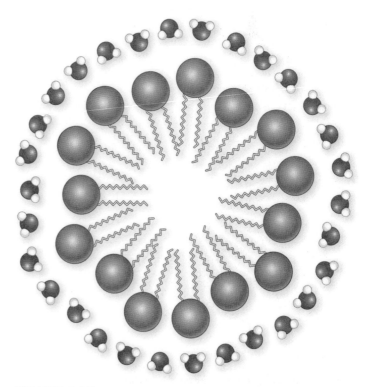

FIGURE 2.23 **The formation of a micelle structure by phospholipids such as lecithin.** The hydrophilic outer layer of the micelle faces the aqueous environment.

Steroids

In terms of structure, **steroids** differ considerably from triglycerides or phospholipids, yet steroids are still included in the lipid category of molecules because they are nonpolar and insoluble in water. All steroid molecules have the same basic structure: three six-carbon rings joined to one five-carbon ring (fig. 2.24). However, different kinds of steroids have different functional groups attached to this basic structure, and they vary in the number and position of the double covalent bonds between the carbon atoms in the rings.

Cholesterol is an important molecule in the body because it serves as the precursor (parent molecule) for the steroid hormones produced by the gonads and adrenal cortex. The testes and ovaries (collectively called the *gonads*) secrete **sex steroids,** which include estradiol and progesterone from the ovaries and testosterone from the testes. The adrenal cortex secretes the **corticosteroids,** including hydrocortisone and aldosterone, as well as weak androgens (including dehydroepiandrosterone, or DHEA). Cholesterol is also an important component of cell membranes, and serves as the precursor molecule for bile salts and vitamin D_3.

Prostaglandins

Prostaglandins are a type of fatty acid with a cyclic hydrocarbon group. Their name is derived from their original discovery in the semen as a secretion of the prostate. However, we now know that they are produced in almost all organs where they serve a variety of regulatory functions. Prostaglandins are implicated in the regulation of blood vessel diameter, ovulation, uterine contraction during labor, inflammation reactions, blood clotting, and many other functions. Structural formulas for different types of prostaglandins are shown in figure 2.25.

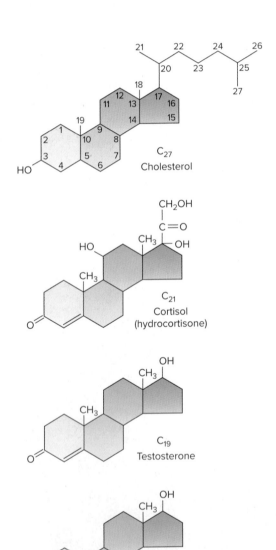

FIGURE 2.24 Cholesterol and some of the steroid hormones derived from cholesterol. The steroid hormones are secreted by the gonads and the adrenal cortex. The carbon atoms in cholesterol are indicated by the numbers.

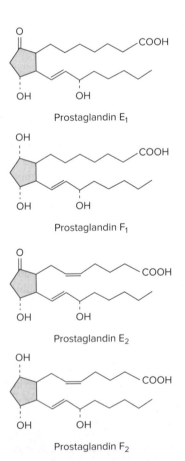

FIGURE 2.25 Structural formulas for various prostaglandins. Prostaglandins are a family of regulatory compounds derived from a membrane lipid known as arachidonic acid.

CHECKPOINTS

5a. Describe the structural characteristic of all carbohydrates, and distinguish between monosaccharides, disaccharides, and polysaccharides.

5b. Describe the characteristics of a lipid, and discuss the different subcategories of lipids.

6. Explain, in terms of dehydration synthesis and hydrolysis reactions, how disaccharides and monosaccharides can be interconverted and how triglycerides can be formed and broken down.

7. Relate the functions of phospholipids to their structure, and explain the significance of the prostaglandins.

2.3 PROTEINS

Proteins are large molecules composed of amino acid subunits. There are about 20 different types of amino acids that can be used in constructing a given protein, so the variety of protein structures is immense. This variety allows each type of protein to perform very specific functions.

LEARNING OUTCOMES

After studying this section, you should be able to:

8. Describe amino acids and explain how peptide bonds between them are formed and broken.

9. Describe the different orders of protein structure, the different functions of proteins, and how protein structure grants specificity of function.

The enormous diversity of protein structure results from the fact that there are 20 different building blocks—the *amino acids*—that can be used to form a protein. These amino acids, as will be described shortly, are joined together to form a chain. Because of chemical interactions between the amino acids, the chain can twist and fold in a specific manner. The sequence of amino acids in a protein, and thus the specific structure of the protein, is determined by genetic information. This genetic information for protein synthesis is contained in another category of organic molecules, the *nucleic acids,* which includes the macromolecules DNA and RNA. The structure of nucleic acids is described in the next section, and the mechanisms by which the genetic information they encode directs protein synthesis are described in chapter 3.

Structure of Proteins

Proteins consist of long chains of subunits called **amino acids.** As the name implies, each amino acid contains an *amino*

group (NH_2) on one end of the molecule and a *carboxyl group* (COOH) on another end. There are about 20 different amino acids, each with a distinct structure and chemical properties, that are used to build proteins. The differences between the amino acids are due to differences in their *functional groups.* R is the abbreviation for the functional group in the general formula for an amino acid (fig. 2.26). The R symbol actually stands for the word *residue,* but it can be thought of as indicating the "*rest of* the molecule."

When amino acids are joined together by dehydration synthesis, the hydrogen from the amino end of one amino acid combines with the hydroxyl group in the carboxyl end of another amino acid. As a covalent bond is formed between the two amino acids, water is produced (fig. 2.27). The bond between adjacent amino acids is called a **peptide bond,** and the compound formed is called a *peptide.* Two amino acids bound together are called a *dipeptide;* three, a *tripeptide.* When numerous amino acids are joined in this way, a chain of amino acids, or a **polypeptide,** is produced.

The lengths of polypeptide chains vary widely. A hormone called *thyrotropin-releasing hormone,* for example, is

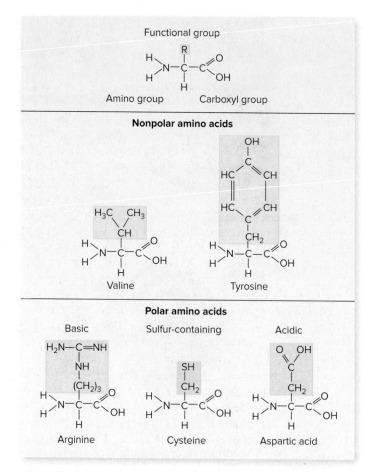

FIGURE 2.26 **Representative amino acids.** The figure depicts different types of functional (R) groups. Each amino acid differs from other amino acids in the number and arrangement of atoms in its functional groups.

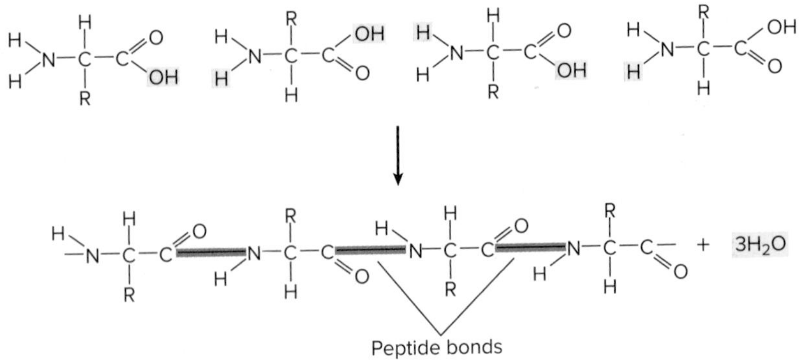

FIGURE 2.27   The formation of peptide bonds by dehydration synthesis reactions. Water molecules are split off as the peptide bonds (highlighted in red) are produced between the amino acids.

only three amino acids long, whereas myosin, a muscle protein, contains about 4,500 amino acids. When the length of a polypeptide chain becomes very long (containing more than about 100 amino acids), the molecule is called a *protein*.

The structure of a protein can be described at four different levels. The first level of structure describes the sequence of amino acids in the particular protein; this is the **primary structure** of the protein. Each type of protein has a different primary structure. All of the billions of *copies* of a given type of protein in a person have the same structure, however, because the primary structure of a given protein is coded by the person's genes. The primary structure of a protein is illustrated in figure 2.28*a*.

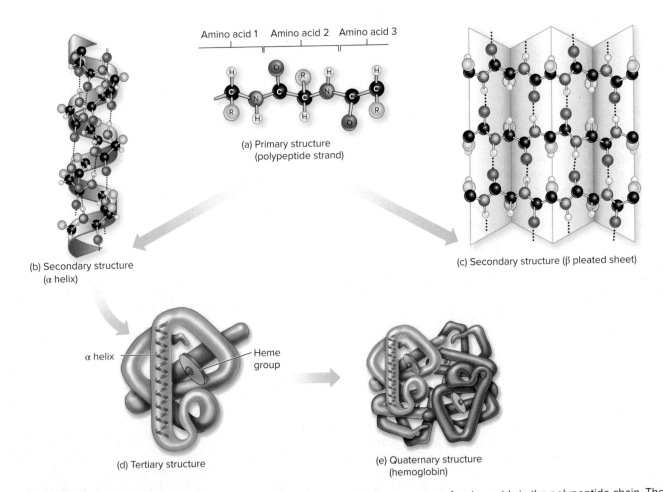

(a) Primary structure (polypeptide strand)

(b) Secondary structure (α helix)

(c) Secondary structure (β pleated sheet)

(d) Tertiary structure

(e) Quaternary structure (hemoglobin)

FIGURE 2.28   The structure of proteins. The primary structure (*a*) is the sequence of amino acids in the polypeptide chain. The secondary structure is the conformation of the chain created by hydrogen bonding between amino acids; this can be either an alpha helix (*b*) or a beta pleated sheet (*c*). The tertiary structure (*d*) is the three-dimensional structure of the protein. The formation of a protein by the bonding together of two or more polypeptide chains is the quaternary structure (*e*) of the protein. Hemoglobin, the protein in red blood cells that carries oxygen, is used here as an example. Each heme group contains a central iron (Fe^{2+}) surrounded by a flat organic molecular structure. **AP|R**

Weak hydrogen bonds may form between the hydrogen atom of an amino group and an oxygen atom from a different amino acid nearby. These weak bonds cause the polypeptide chain to assume a particular shape, known as the **secondary structure** of the protein (fig. 2.28b,c). This can be the shape of an *alpha (α) helix,* or alternatively, the shape of what is called a *beta (β) pleated sheet.*

Most polypeptide chains bend and fold upon themselves to produce complex three-dimensional shapes called the **tertiary structure** of the protein (fig. 2.28d). Each type of protein has its own characteristic tertiary structure. This is because the folding and bending of the polypeptide chain is produced by chemical interactions between particular amino acids located in different regions of the chain.

Most of the tertiary structure of proteins is formed and stabilized by weak chemical interactions between the functional groups of amino acids located some distance apart along the polypeptide chain. In terms of their strengths, these weak interactions are relatively stronger for ionic bonds, weaker for hydrogen bonds, and weakest for van der Waals forces (fig. 2.29). The natures of ionic bonds and hydrogen bonds have been previously discussed. *Van der Waals forces are weak forces between electrically neutral molecules that come very close together.* These forces occur because, even in electrically neutral molecules, the electrons are not always evenly distributed but can at some instants be found at one end of the molecule.

Because most of the tertiary structure is stabilized by weak bonds, this structure can easily be disrupted by high temperature or by changes in pH. Changes in the tertiary structure of proteins that occur by these means are referred to as *denaturation* of the proteins. The tertiary structure of some proteins, however, is made more stable by strong covalent bonds between sulfur atoms (called *disulfide bonds* and abbreviated S—S) in the functional group of an amino acid known as cysteine (fig. 2.29).

Denatured proteins retain their primary structure (the peptide bonds are not broken) but have altered chemical properties. Cooking a pot roast, for example, alters the texture of the meat proteins—it doesn't result in an amino acid soup. Denaturation is most dramatically demonstrated by frying an egg. Egg albumin proteins are soluble in their native state in which they form the clear, viscous fluid of a raw egg. When

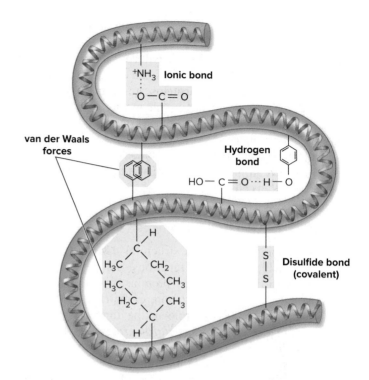

FIGURE 2.29 **The bonds responsible for the tertiary structure of a protein.** The tertiary structure of a protein is held in place by a variety of bonds. These include relatively weak bonds, such as hydrogen bonds, ionic bonds, and van der Waals (hydrophobic) forces, as well as the strong covalent disulfide bonds.

denatured by cooking, these proteins change shape, cross-bond with each other, and by this means form an insoluble white precipitate—the egg white.

Hemoglobin and insulin are composed of a number of polypeptide chains covalently bonded together. This is the **quaternary structure** of these molecules. Insulin, for example, is composed of two polypeptide chains—one that is 21 amino acids long, the other that is 30 amino acids long. Hemoglobin (the protein in red blood cells that carries oxygen) is composed of four separate polypeptide chains (see fig. 2.28e). The composition of various body proteins is shown in table 2.4.

TABLE 2.4 | Composition of Selected Proteins Found in the Body

Protein	Number of Polypeptide Chains	Nonprotein Component	Function
Hemoglobin	4	Heme pigment	Carries oxygen in the blood
Myoglobin	1	Heme pigment	Stores oxygen in muscle
Insulin	2	None	Hormonal regulation of metabolism
Blood group proteins	1	Carbohydrate	Produces blood types
Lipoproteins	1	Lipids	Transports lipids in blood

Many proteins in the body are normally found combined, or *conjugated*, with other types of molecules. **Glycoproteins** are proteins conjugated with carbohydrates. Examples of such molecules include certain hormones and some proteins found in the cell membrane. **Lipoproteins** are proteins conjugated with lipids. These are found in cell membranes and in the plasma (the fluid portion of the blood). Proteins may also be conjugated with pigment molecules. These include hemoglobin, which transports oxygen in red blood cells, and the cytochromes, which are needed for oxygen utilization and energy production within cells.

Functions of Proteins

Because of their tremendous structural diversity, proteins can serve a wider variety of functions than any other type of molecule in the body. Many proteins, for example, contribute significantly to the structure of different tissues and in this way play a passive role in the functions of these tissues. Examples of such *structural proteins* include collagen (fig. 2.30) and keratin. Collagen is a fibrous protein that provides tensile strength to connective tissues, such as tendons and ligaments. Keratin is found in the outer layer of dead cells in the epidermis, where it prevents water loss through the skin.

Many proteins play a more active role in the body where specificity of structure and function is required. *Enzymes* and *antibodies,* for example, are proteins—no other type of molecule could provide the vast array of different structures needed for their tremendously varied functions. As another example, proteins in cell membranes may serve as *receptors* for specific regulatory molecules (such as hormones) and as *carriers* for transport of specific molecules across the membrane. Proteins provide the diversity of shape and chemical properties required by these functions.

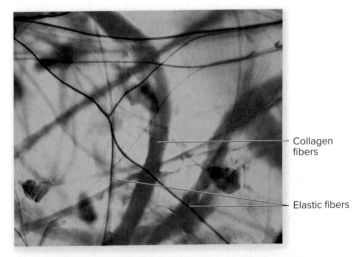

FIGURE 2.30   **A photomicrograph of collagen fibers within connective tissue.** Collagen proteins strengthen the connective tissues. **AP|R** © Ed Reschke

Collagen fibers

Elastic fibers

2.4 NUCLEIC ACIDS

Nucleic acids include the macromolecules DNA and RNA, which are critically important in genetic regulation, and the subunits from which these molecules are formed. These subunits are known as nucleotides.

LEARNING OUTCOMES

After studying this section, you should be able to:

10. Describe the structure of nucleotides and distinguish between the structure of DNA and RNA.

11. Explain the law of complementary base pairing, and describe how that occurs between the two strands of DNA.

Nucleotides are the subunits of nucleic acids, bonded together in dehydration synthesis reactions to form long polynucleotide chains. Each nucleotide, however, is itself composed of three smaller subunits: a five-carbon (*pentose*) sugar, a phosphate group attached to one end of the sugar, and a *nitrogenous base* attached to the other end of the sugar (fig. 2.31). The nitrogenous bases are nitrogen-containing molecules of two kinds: pyrimidines and purines. The *pyrimidines* contain a single ring of carbon and nitrogen, whereas the *purines* have two such rings.

Deoxyribonucleic Acid

The structure of **DNA (deoxyribonucleic acid)** serves as the basis for the genetic code. For this reason, it might seem logical that DNA should have an extremely complex structure. DNA is indeed larger than any other molecule in the cell, but its structure is actually simpler than that of most proteins. This simplicity of structure deceived some early investigators into believing that the protein content of chromosomes, rather than their DNA content, provided the basis for the genetic code.

Sugar molecules in the nucleotides of DNA are a type of pentose (five-carbon) sugar called **deoxyribose.** Each deoxyribose can be covalently bonded to one of four possible bases. These bases include the two purines (**guanine** and **adenine**) and

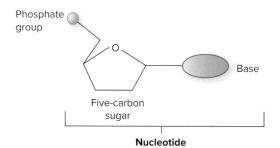

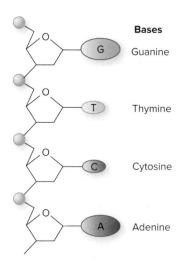

FIGURE 2.31 **The structure of nucleic acids.** The components of a single nucleotide (*above*), and the structure of a polynucleotide (*below*). The polynucleotide was formed by dehydration synthesis reactions between nucleotides that join the nucleotides together by sugar-phosphate bonds. **AP|R**

the two pyrimidines (**cytosine** and **thymine**) (fig. 2.32). There are thus four different types of nucleotides that can be used to produce the long DNA chains. If you remember that there are about 20 different amino acids used to produce proteins, you can now understand why many scientists were deceived into thinking that genes were composed of proteins rather than nucleic acids.

When nucleotides combine to form a chain, the phosphate group of one condenses with the deoxyribose sugar of another nucleotide. This forms a sugar-phosphate chain as water is removed in dehydration synthesis. Because the nitrogenous bases are attached to the sugar molecules, the sugar-phosphate chain looks like a "backbone" from which the bases project. Each of these bases can form hydrogen bonds with other bases, which are in turn joined to a different chain of nucleotides. Such hydrogen bonding between bases thus produces a *double-stranded* DNA molecule; the two strands are like a staircase, with the paired bases as steps (fig. 2.33).

Actually, the two chains of DNA twist about each other to form a **double helix,** so that the molecule resembles a spiral staircase (fig. 2.33). It has been shown that the number of purine bases in DNA is equal to the number of pyrimidine bases. The reason for this is explained by the **law of complementary base pairing:** *adenine can pair only with thymine* (through two hydrogen bonds), whereas *guanine can pair only with cytosine* (through three hydrogen bonds). With knowledge of this rule, we could predict the base sequence

FIGURE 2.32 **The four nitrogenous bases in deoxyribonucleic acid (DNA).** Notice that hydrogen bonds can form between guanine and cytosine and between thymine and adenine. **AP|R**

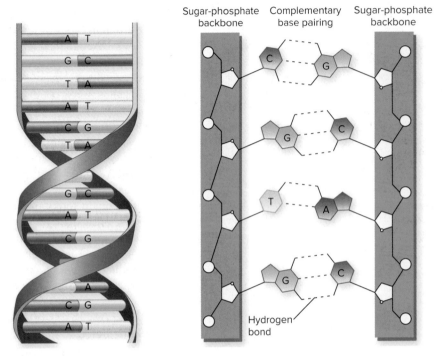

FIGURE 2.33   The double-helix structure of DNA. The two strands are held together by hydrogen bonds between complementary bases in each strand. AP|R

of one DNA strand if we knew the sequence of bases in the complementary strand.

Although we can be certain which base is opposite a given base in DNA, we cannot predict which bases will be above or below that particular pair within a single polynucleotide chain. Although there are only four bases, the number of possible base sequences along a stretch of several thousand nucleotides (the length of most genes) is almost infinite. To gain perspective, it is useful to realize that the total human *genome* (all of the genes in a cell) consists of over 3 billion base pairs that would extend over a meter if the DNA molecules were unraveled and stretched out.

Yet even with this amazing variety of possible base sequences, almost all of the billions of copies of a particular gene in a person are identical. The mechanisms by which identical DNA copies are made and distributed to the daughter cells when a cell divides will be described in chapter 3.

Ribonucleic Acid

DNA can direct the activities of the cell only by means of another type of nucleic acid—**RNA (ribonucleic acid).** Like DNA, RNA consists of long chains of nucleotides joined together by sugar-phosphate bonds. Nucleotides in RNA, however, differ from those in DNA (fig. 2.34) in three ways: (1) a **ribonucleotide** contains the sugar **ribose** (instead of deoxyribose), (2) the base **uracil** is found in place of thymine, and (3) RNA is composed of a single polynucleotide strand (it is not double-stranded like DNA). The structure and function of RNA are described in chapter 3, section 3.3.

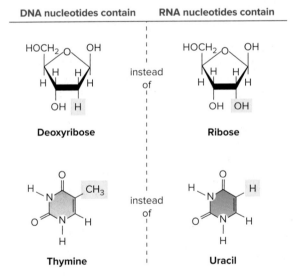

FIGURE 2.34   Differences between the nucleotides and sugars in DNA and RNA. DNA has deoxyribose and thymine; RNA has ribose and uracil. The other three bases are the same in DNA and RNA. AP|R

There are three major types of RNA molecules that function in the cytoplasm of cells: *messenger RNA (mRNA), transfer RNA (tRNA),* and *ribosomal RNA (rRNA).* All three types are made within the cell nucleus by using information contained in DNA as a guide. The functions of RNA are described in chapter 3.

In addition to their participation in genetic regulation as part of RNA, purine-containing nucleotides are used for other purposes as well. These include roles as energy carriers (ATP and GTP); regulation of cellular events (cyclic AMP, or cAMP); and coenzymes (nicotinamide adenine dinucleotide, or NAD; and flavine adenine dinucleotide, or FAD). These are discussed in chapters 4, 5, and 6. Purines (ATP and adenosine) are even used as neurotransmitters by some neurons (chapter 7, section 7.6).

CHECKPOINTS

10a. What are nucleotides, and of what are they composed?

10b. List the types of RNA, and explain how the structure of RNA differs from the structure of DNA.

11. Describe the structure of DNA, and explain the law of complementary base pairing.

CLINICAL INVESTIGATION   SUMMARY

Brian has multiple myeloma, a type of cancer affecting cells of the immune system called plasma cells, which secrete antibodies. This disease has different degrees of severity and various forms of treatment, including the use of thalidomide. Brian needn't be concerned about the teratogenic effects of one of the enantiomers of thalidomide, because he obviously won't become pregnant. His ketonuria is related to his weight loss, because ketone bodies are produced in the liver from fatty acids released by adipose cells when they hydrolyze their stored triglycerides.

See the additional chapter 2 Clinical Investigation on *High Cholesterol* in the Connect site for this text.

SUMMARY

2.1 Atoms, Ions, and Chemical Bonds   25

A. Covalent bonds are formed by atoms that share electrons. They are the strongest type of chemical bond.
  1. Electrons are equally shared in nonpolar covalent bonds and unequally shared in polar covalent bonds.
  2. Atoms of oxygen, nitrogen, and phosphorus strongly attract electrons and become electrically negative compared to the other atoms sharing electrons with them.

B. Ionic bonds are formed by atoms that transfer electrons. These weak bonds join atoms together in an ionic compound.
  1. If one atom in this compound takes an electron from another atom, it gains a net negative charge and the other atom becomes positively charged.
  2. Ionic bonds easily break when the ionic compound is dissolved in water. Dissociation of the ionic compound yields charged atoms called ions.

C. When hydrogen bonds with an electronegative atom, it gains a slight positive charge and is weakly attracted to another electronegative atom. This weak attraction is a hydrogen bond.

D. Acids donate hydrogen ions to solution, whereas bases lower the hydrogen ion concentration of a solution.
  1. The pH scale is a negative function of the logarithm of the hydrogen ion concentration.
  2. In a neutral solution, the concentration of H^+ is equal to the concentration of OH^-, and the pH is 7.
  3. Acids raise the H^+ concentration and thus lower the pH below 7; bases lower the H^+ concentration and thus raise the pH above 7.

E. Organic molecules contain atoms of carbon and hydrogen joined together by covalent bonds. Atoms of nitrogen, oxygen, phosphorus, or sulfur may be present as specific functional groups in the organic molecule.

2.2 Carbohydrates and Lipids   33

A. Carbohydrates contain carbon, hydrogen, and oxygen, usually in a ratio of 1:2:1.
  1. Carbohydrates consist of simple sugars (monosaccharides), disaccharides, and polysaccharides (such as glycogen).
  2. Covalent bonds between monosaccharides are formed by dehydration synthesis, or condensation. Bonds are broken by hydrolysis reactions.

B. Lipids are organic molecules that are insoluble in polar solvents such as water.
  1. Triglycerides (fat and oil) consist of three fatty acid molecules joined to a molecule of glycerol.
  2. Ketone bodies are smaller derivatives of fatty acids that can be used for energy by some organs, such as the brain.
  3. Phospholipids (such as lecithin) are phosphate-containing lipids that have a hydrophilic polar group. The rest of the molecule is hydrophobic.
  4. Steroids (including the hormones of the adrenal cortex and gonads) are lipids with a characteristic four-ring structure.
  5. Prostaglandins are a family of cyclic fatty acids that serve a variety of regulatory functions.

2.3 Proteins   41

A. Proteins are composed of long chains of amino acids bound together by covalent peptide bonds.
  1. Each amino acid contains an amino group, a carboxyl group, and a functional group. Differences in the functional groups give each of the more than 20 different amino acids an individual identity.
  2. The polypeptide chain may be twisted into a helix (secondary structure) and bent and folded to form the tertiary structure of the protein.

3. Proteins that are composed of two or more polypeptide chains are said to have a quaternary structure.

4. Proteins may be combined with carbohydrates, lipids, or other molecules.

5. Because they are so diverse structurally, proteins serve a wider variety of specific functions than any other type of molecule.

2.4 Nucleic Acids   44

A. DNA is composed of four nucleotides, each of which contains the sugar deoxyribose.

1. Two of the bases contain the purines adenine and guanine; two contain the pyrimidines cytosine and thymine.

2. DNA consists of two polynucleotide chains joined together by hydrogen bonds between their bases.

3. Hydrogen bonds can only form between the bases adenine and thymine, and between the bases guanine and cytosine.

4. This complementary base pairing is critical for DNA synthesis and for genetic expression.

B. RNA consists of four nucleotides, each of which contains the sugar ribose.

1. The nucleotide bases are adenine, guanine, cytosine, and uracil (in place of the DNA base thymine).

2. RNA consists of only a single polynucleotide chain.

3. There are different types of RNA, which have different functions in genetic expression.

REVIEW ACTIVITIES

Test Your Knowledge

1. Which of these statements about atoms is *true?*
   a. They have more protons than electrons.
   b. They have more electrons than protons.
   c. They are electrically neutral.
   d. They have as many neutrons as they have electrons.

2. The bond between oxygen and hydrogen in a water molecule is
   a. a hydrogen bond.
   b. a polar covalent bond.
   c. a nonpolar covalent bond.
   d. an ionic bond.

3. Which of these is a nonpolar covalent bond?
   a. bond between two carbons
   b. bond between sodium and chloride
   c. bond between two water molecules
   d. bond between nitrogen and hydrogen

4. Solution A has a pH of 2, and solution B has a pH of 10. Which of these statements about these solutions is *true?*
   a. Solution A has a higher H^+ concentration than solution B.
   b. Solution B is basic.
   c. Solution A is acidic.
   d. All of these are true.

5. Glucose is
   a. a disaccharide.   c. a monosaccharide.
   b. a polysaccharide.   d. a phospholipid.

6. Digestion reactions occur by means of
   a. dehydration synthesis.
   b. hydrolysis.

7. Carbohydrates are stored in the liver and muscles in the form of
   a. glucose.   c. glycogen.
   b. triglycerides.   d. cholesterol.

8. Lecithin is
   a. a carbohydrate.
   b. a protein.
   c. a steroid.
   d. a phospholipid.

9. Which of these lipids have regulatory roles in the body?
   a. steroids
   b. prostaglandins
   c. triglycerides
   d. both *a* and *b*
   e. both *b* and *c*

10. The tertiary structure of a protein is *directly* determined by
    a. genes.
    b. the primary structure of the protein.
    c. enzymes that "mold" the shape of the protein.
    d. the position of peptide bonds.

11. The type of bond formed between two molecules of water is
    a. a hydrolytic bond.
    b. a polar covalent bond.
    c. a nonpolar covalent bond.
    d. a hydrogen bond.

12. The carbon-to-nitrogen bond that joins amino acids together is called
    a. a glycosidic bond.
    b. a peptide bond.
    c. a hydrogen bond.
    d. a double bond.

13. The RNA nucleotide base that pairs with adenine in DNA is
    a. thymine.   c. guanine.
    b. uracil.   d. cytosine.

14. If four bases in one DNA strand are A (adenine), G (guanine), C (cytosine), and T (thymine), the complementary bases in the RNA strand made from this region are

    a. T,C,G,A.   **c.** A,G,C,U.
    b. C,G,A,U.   **d.** U,C,G,A.

15. The type of bonds between the nucleotide bases of the two strands of DNA are

    a. polar covalent bonds.

    b. nonpolar covalent bonds.

    c. hydrogen bonds.

    d. ionic bonds.

Test Your Understanding

16. Compare and contrast nonpolar covalent bonds, polar covalent bonds, and ionic bonds.

17. Define *acid* and *base* and explain how acids and bases influence the pH of a solution.

18. Explain, in terms of dehydration synthesis and hydrolysis reactions, the relationships between starch in an ingested potato, liver glycogen, and blood glucose.

19. "All fats are lipids, but not all lipids are fats." Explain why this is an accurate statement.

20. What are the similarities and differences between a fat and an oil? Comment on the physiological and clinical significance of the degree of saturation of fatty acid chains.

21. Explain how one DNA molecule serves as a template for the formation of another DNA molecule and why DNA synthesis is said to be semiconservative.

Test Your Analytical Ability

22. Explain the relationship between the primary structure of a protein and its secondary and tertiary structures. What do you think would happen to the tertiary structure if some amino acids were substituted for others in the primary structure? What physiological significance might this have?

23. Suppose you try to discover a hormone by homogenizing an organ in a fluid, filtering the fluid to eliminate the solid material, and then injecting the extract into an animal to see the effect. If an aqueous (water) extract does not work, but one using benzene as the solvent does have an effect, what might you conclude about the chemical nature of the hormone? Explain.

24. From the ingredients listed on a food wrapper, it would appear that the food contains high amounts of fat. Yet on the front of the package is the large slogan, "Cholesterol Free!" In what sense is this slogan chemically correct? In what way is it misleading?

25. A butter substitute says "Nonhydrogenated, zero trans fats" on the label. Explain the meaning of these terms and their relationship to health.

26. When you cook a pot roast, you don't end up with an amino acid soup. Explain why this is true, in terms of the strengths of the different types of bonds in a protein.

Test Your Quantitative Ability

The molecular weight is the sum of the atomic weights (mass numbers) of its atoms. Use table 2.1 to perform the following calculations.

27. Calculate the molecular weight of water (H_2O) and glucose ($C_6H_{12}O_6$).

28. Given that fructose is a structural isomer of glucose (see fig. 2.14), what is its molecular weight?

29. Review the dehydration synthesis of sucrose in figure 2.16*b* and calculate the molecular weight of sucrose.

30. Account for the difference between the molecular weight of sucrose and the sum of the molecular weights of glucose and fructose.

ONLINE STUDY TOOLS

 |

3

Cell Structure and Genetic Control

CLINICAL INVESTIGATION

George, who is 28 years old, complains of pain in his hips and knees, and he has a swollen abdomen. Upon extensive medical examination, he is found to have hepatomegaly (an enlarged liver) and splenomegaly (an enlarged spleen). He thinks his enlarged liver might be due to his abuse of alcohol and drugs.

Some of the new terms and concepts you will encounter include:

- Lysosomes and lysosomal storage diseases
- Rough and smooth endoplasmic reticulum

3.1 PLASMA MEMBRANE AND ASSOCIATED STRUCTURES

The cell is the basic unit of structure and function in the body. Many of the functions of cells are performed by particular subcellular structures known as organelles. The plasma (cell) membrane allows selective communication between the intracellular and extracellular compartments and aids cellular movement.

LEARNING OUTCOMES

After studying this section, you should be able to:

1. Describe the structure of the plasma membrane, cilia, and flagella.
2. Describe amoeboid movement, phagocytosis, pinocytosis, receptor-mediated endocytosis, and exocytosis.

Cells look so small and simple when viewed with the ordinary (light) microscope that it is difficult to think of each one as a living entity unto itself. Equally amazing is the fact that the physiology of our organs and systems derives from the complex functions of the cells of which they are composed. Complexity of function demands complexity of structure, even at the subcellular level.

As the basic functional unit of the body, each cell is a highly organized molecular factory. Cells come in a wide variety of shapes and sizes. This great diversity, which is also apparent in the subcellular structures within different cells, reflects the diversity of function of different cells in the body. All cells, however, share certain characteristics; for example, they are all surrounded by a plasma membrane, and most of them possess the structures listed in table 3.1. Thus, although no single cell can be considered "typical," the general structure of cells can be indicated by a single illustration (fig. 3.1).

For descriptive purposes, a cell can be divided into three principal parts:

1. **Plasma (cell) membrane.** The selectively permeable plasma membrane surrounds the cell, gives it form, and separates the cell's internal structures from the extracellular environment. The plasma membrane also participates in intercellular communication.

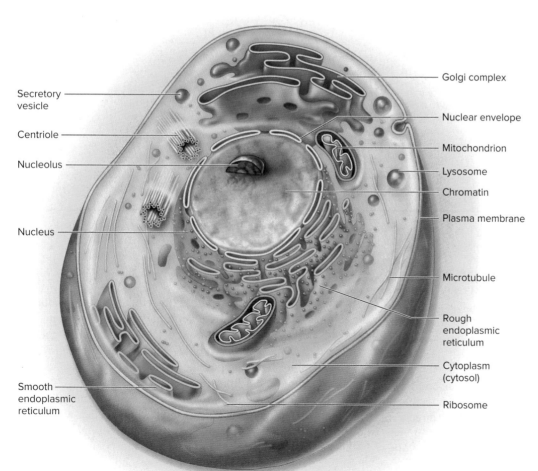

Secretory vesicle

Centriole

Nucleolus

Nucleus

Smooth endoplasmic reticulum

Golgi complex

Nuclear envelope

Mitochondrion

Lysosome

Chromatin

Plasma membrane

Microtubule

Rough endoplasmic reticulum

Cytoplasm (cytosol)

Ribosome

FIGURE 3.1   A generalized human cell showing the principal organelles. Because most cells of the body are highly specialized, they have structures that differ from those shown here. AP|R

TABLE 3.1 | Cellular Components: Structure and Function

Component	Structure	Function
Plasma (cell) membrane	Membrane composed of double layer of phospholipids in which proteins are embedded	Gives form to cell and controls passage of materials into and out of cell
Cytoplasm	Fluid, jellylike substance between the plasma membrane and the nucleus in which organelles are suspended	Serves as matrix substance in which chemical reactions occur
Endoplasmic reticulum	System of interconnected membrane-forming canals and tubules	Smooth endoplasmic reticulum metabolizes nonpolar compounds and stores Ca^{2+} in striated muscle cells, rough endoplasmic reticulum assists in protein synthesis
Ribosomes	Granular particles composed of protein and RNA	Synthesize proteins
Golgi complex	Cluster of flattened membranous sacs	Synthesizes carbohydrates and packages molecules from the endoplasmic reticulum for secretion; secretes lipids and glycoproteins
Mitochondria	Membranous sacs with folded inner partitions	Release energy from food molecules and transform energy into usable ATP
Lysosomes	Membranous sacs	Digest foreign molecules and worn and damaged organelles
Peroxisomes	Spherical membranous vesicles	Contain enzymes that detoxify harmful molecules and break down hydrogen peroxide
Centrosome	Nonmembranous mass of two rodlike centrioles	Helps to organize spindle fibers and distribute chromosomes during mitosis
Vacuoles	Membranous sacs	Store and release various substances within the cytoplasm
Microfilaments and microtubules	Thin, hollow tubes	Support cytoplasm and transport materials within the cytoplasm
Cilia and flagella	Minute cytoplasmic projections that extend from the cell surface	Move particles along cell surface or move the cell
Nuclear envelope	Double-layered membrane that surrounds the nucleus, composed of protein and lipid molecules	Supports nucleus and controls passage of materials between nucleus and cytoplasm
Nucleolus	Dense nonmembranous mass composed of protein and RNA molecules	Produces ribosomal RNA for ribosomes
Chromatin	Fibrous strands composed of protein and DNA	Contains genetic code that determines which proteins (including enzymes) will be manufactured by the cell

2. **Cytoplasm and organelles.** The cytoplasm is the aqueous content of a cell inside the plasma membrane but outside the nucleus. Organelles (excluding the nucleus) are subcellular structures within the cytoplasm that perform specific functions. The term **cytosol** is frequently used to describe the fluid portion of the cytoplasm—that is, the part that cannot be removed by centrifugation.

3. **Nucleus.** The nucleus is a large, generally spheroid body within a cell. The largest of the organelles, it contains the DNA, or genetic material, of the cell and thus directs the cell's activities. The nucleus also contains one or more *nucleoli*. Nucleoli are centers for the production of ribosomes, which are the sites of protein synthesis.

Structure of the Plasma Membrane

Because the intracellular and extracellular environments (or "compartments") are both aqueous, a barrier must be present to prevent the loss of enzymes, nucleotides, and other cellular molecules that are water-soluble. This barrier surrounding the cell cannot itself be composed of water-soluble molecules; it is instead composed of lipids.

The **plasma membrane** (also called the **cell membrane**), like all of the membranes surrounding organelles within the cell, is composed primarily of phospholipids and proteins. Phospholipids, described in chapter 2, are polar (and hydrophilic) in the region that contains the phosphate group and nonpolar (and hydrophobic) throughout the rest of the molecule. Because the environment on each side of the membrane is aqueous, the hydrophobic parts of the molecules "huddle together" in the center of the membrane, leaving the polar parts exposed to water on both surfaces. This results in the formation of a double layer of phospholipids in the plasma membrane.

The hydrophobic middle of the membrane restricts the passage of water and water-soluble molecules and ions. Certain of these polar compounds, however, do pass through the membrane. The specialized functions and selective transport properties of the membrane are primarily due to its protein content. Membrane proteins are described as peripheral or integral. *Peripheral proteins* are only partially embedded in one face of the membrane, whereas *integral proteins* span the membrane from one side to the other. Because the membrane is not solid—phospholipids and proteins are free to move laterally— the proteins within the phospholipid "sea" are not uniformly distributed. Rather, they present a constantly changing mosaic pattern, an arrangement known as the **fluid-mosaic model** of membrane structure (fig. 3.2).

Scientists now recognize that the fluid-mosaic model of the plasma membrane is somewhat misleading, in that the membrane is not as uniform in structure as implied by figure 3.2. The proteins in the plasma membrane can be localized according to their function, so that their distribution is patchy rather than uniform. Thus, proteins in some regions are much more crowded together in the plasma membrane than is indicated in figure 3.2. This can help the proteins better perform their functions.

The proteins in the plasma membrane serve a variety of functions. These include:

- structural proteins that connect both to the cytoskeleton (section 3.2) and to extracellular proteins;
- enzyme proteins that are specific for particular molecules so that they can catalyze specific reactions (chapter 4);
- receptor proteins for specific extracellular chemical regulators, including particular neurotransmitters and hormones (chapter 6);
- specific antigen "markers" that identify the cell to the immune system (chapter 15).

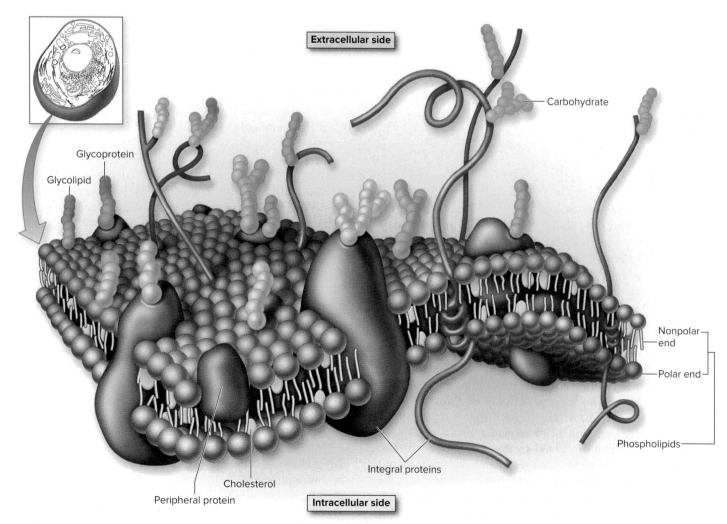

FIGURE 3.2 **The fluid-mosaic model of the plasma membrane.** The membrane consists of a double layer of phospholipids, with the polar regions (shown by spheres) oriented outward and the nonpolar hydrocarbons (wavy tails) oriented toward the center. Proteins may completely or partially span the membrane. Carbohydrates are attached to the outer surface. AP|R

In addition to lipids and proteins, the plasma membrane also contains carbohydrates, which are primarily attached to the outer surface of the membrane as glycoproteins and glycolipids. Certain glycolipids on the plasma membrane of red blood cells serve as antigens that determine the blood type. Other carbohydrates on the plasma membrane have numerous negative charges and, as a result, affect the interaction of regulatory molecules with the membrane. The negative charges at the surface also affect interactions between cells—they help keep red blood cells apart, for example. Stripping the carbohydrates from the outer surface of red blood cells results in the cells being destroyed more rapidly by the liver, spleen, and bone marrow.

CLINICAL APPLICATION

The cholesterol content of plasma membranes (generally 20% to 25% of total membrane lipids) contributes to their flexibility, and an inherited defect in this ratio can cause the red blood cells to be unable to flex as they pass through narrow blood vessels. Disorders of the protein content of the plasma membrane depend upon the function of the protein. **Cystic fibrosis,** for example, is produced by a defect in a specific ion channel protein; **Duchenne muscular dystrophy** results when the lack of a plasma membrane protein called *dystrophin* prevents fibers of the cytoskeleton from attaching and providing needed support to the plasma membrane. Also, inappropriate enzyme activity associated with the plasma membrane can produce cellular proteins that may contribute to **Alzheimer's disease.**

Phagocytosis

Most of the movement of molecules and ions between the intracellular and extracellular compartments involves passage through the plasma membrane (chapter 6). However, the plasma membrane also participates in the **bulk transport** of larger portions of the extracellular environment. Bulk transport includes the processes of *phagocytosis* and *endocytosis.*

White blood cells known as *neutrophils,* and connective tissue cells called *macrophages* (literally, "big eaters"), are able to perform **amoeboid movement** (move like an amoeba, a single-celled animal). This involves extending parts of their cytoplasm to form *pseudopods* (false feet), which pull the cell through the *extracellular matrix*—generally, an extracellular gel of proteins and carbohydrates. This process depends on the bonding of proteins called *integrins,* which span the plasma membrane of these cells, with proteins in the extracellular matrix.

Cells that exhibit amoeboid motion—as well as certain liver cells, which are not mobile—use pseudopods to surround and engulf particles of organic matter (such as bacteria). This process is a type of cellular "eating" called **phagocytosis.** It serves to protect the body from invading microorganisms and to remove extracellular debris.

Phagocytic cells surround their victim with pseudopods, which join together and fuse (fig. 3.3). After the inner

Pseudopods forming food vacuole

FIGURE 3.3   Colored scanning electron micrograph of phagocytosis. The phagocytic tissue macrophage is engulfing tuberculosis bacteria (pink) with pseudopods. The pseudopods will fuse so that the bacteria will be inside the cell within an enclosed vacuole. **AP|R** © SPL/Science Source

membrane of the pseudopods has become a continuous membrane surrounding the ingested particle, it pinches off from the plasma membrane. The ingested particle is now contained in an organelle called a *food vacuole* within the cell. The food vacuole will subsequently fuse with an organelle called a lysosome (described later), and the particle will be digested by lysosomal enzymes.

Phagocytosis, largely by neutrophils and macrophages, is an important immune process that defends the body and promotes inflammation. Phagocytosis by macrophages is also needed for the removal of senescent (aged) cells and those that die by *apoptosis* (cell suicide). Phagocytes recognize "eat me" signals—primarily phosphatidylserine—on the plasma membrane surface of dying cells. Apoptosis is a normal, ongoing activity in the body and is not accompanied by inflammation.

Endocytosis

Endocytosis is a process in which the plasma membrane furrows inward, instead of extending outward with pseudopods. One form of endocytosis, **pinocytosis,** is a nonspecific process performed by many cells. The plasma membrane invaginates to produce a deep, narrow furrow. The membrane at the top of this furrow then fuses, and a small vesicle containing the extracellular fluid is pinched off and enters the cell. Pinocytosis allows a cell to engulf large molecules such as proteins, as well as any other molecules that may be present in the extracellular fluid.

Another type of endocytosis involves a smaller area of plasma membrane, and it occurs only in response to specific molecules in the extracellular environment. Because the extracellular molecules must bind to very specific *receptor proteins*

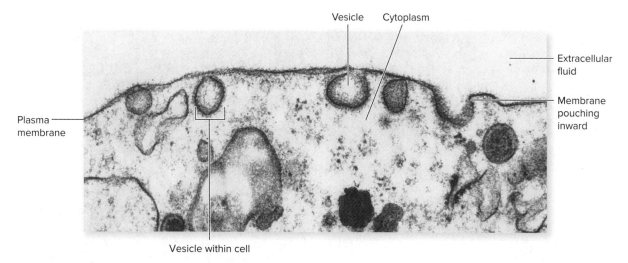

Vesicle   Cytoplasm

Extracellular
fluid

Membrane
pouching
inward

Plasma
membrane

Vesicle within cell

FIGURE 3.4   **Electron micrograph showing endocytosis by a liver cell.** The plasma membrane can be seen to invaginate and create a vesicle that pinches off, containing extracellular material. **AP|R** © Don W. Fawcett/Science Source

in the plasma membrane, this process is known as **receptor-mediated endocytosis.**

In receptor-mediated endocytosis, the interaction of specific molecules in the extracellular fluid with specific membrane receptor proteins causes the membrane to invaginate, fuse, and pinch off to form a vesicle (fig. 3.4). Vesicles formed in this way contain extracellular fluid and molecules that could not have passed by other means into the cell. Cholesterol attached to specific proteins, for example, is taken up into artery cells by receptor-mediated endocytosis. This is in part responsible for atherosclerosis (chapter 13, section 13.7). Hepatitis, polio, and AIDS viruses also exploit the process of receptor-mediated endocytosis to invade cells.

Exocytosis

Exocytosis is a process by which cellular products are secreted into the extracellular environment. Proteins and other molecules produced within the cell that are destined for export (secretion) are packaged within vesicles by an organelle known as the Golgi complex. In the process of exocytosis, these secretory vesicles fuse with the plasma membrane and release their contents into the extracellular environment (see fig. 3.12). Nerve endings, for example, release their chemical neurotransmitters in this manner (chapter 7, section 7.3).

When the vesicle containing the secretory products of the cell fuses with the plasma membrane during exocytosis, the total surface area of the plasma membrane is increased. This process replaces material that was lost from the plasma membrane during endocytosis.

Cilia and Flagella

Cilia are tiny hairlike structures that project from the surface of a cell into the extracellular fluid. *Motile cilia* (those able to move) can beat like rowers in a boat, stroking in unison. Such motile cilia are found in only particular locations in the human body, where they project from the apical surface of epithelial cells (the surface facing the lumen, or cavity) that are stationary and line certain hollow organs. For example, ciliated epithelial cells are found in the respiratory system and the female reproductive tract. In the respiratory airways, the cilia transport strands of mucus to the pharynx (throat), where the mucus can be swallowed or expectorated. In the female reproductive tract, the beating of cilia on the epithelial lining of the uterine tube draws the ovum (egg) into the tube and moves it toward the uterus.

Many epithelial cells that line tubes carrying fluid in the body have a single, nonmotile **primary cilium** on their apical membrane. Movement of fluid bends the primary cilium, resulting in the inflow of Ca^{2+} from the extracellular fluid and the subsequent increase of Ca^{2+} in the cytoplasm. This occurs, for example, in the kidney tubules (chapter 17), as well as in the bile and pancreatic ducts. These events enable the cells to sense fluid movement, although the significance of this sensory function is not yet fully understood.

Cilia are composed of *microtubules* (thin cylinders formed from proteins) and are surrounded by a specialized part of the plasma membrane. There are 9 pairs of microtubules arranged around the circumference of the cilium; in motile cilia, there is also a pair of microtubules in the center, producing an arrangement described as "9 + 2" (fig. 3.5). The nonmotile primary cilium lacks the central pair of microtubules, and so is described as having a "9 + 0" arrangement. Scientists recently discovered that there is a reason microtubules are paired in a cilium: they serve as "rails" in a system in which the paired microtubules transport building materials in opposite directions.

Within the cell cytoplasm at the base of each cilium is a pair of structures called *centrioles,* composed of microtubules and oriented at right angles to each other (see fig. 3.28). The pair form a structure called a *centrosome.* The centriole that points along the axis of the cilium is also known as the *basal body,* and this structure is required to form the

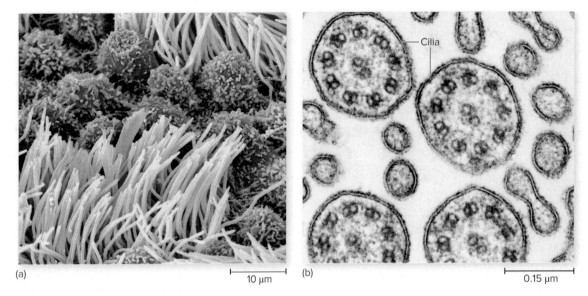

(a)   10 µm   (b)   0.15 µm

FIGURE 3.5   **Cilia, as seen with the electron microscope.**   (*a*) Scanning electron micrograph of cilia on the epithelium lining the trachea; (*b*) transmission electron micrograph of a cross section of cilia, showing the "9 + 2" arrangement of microtubules within each cilium.   AP|R   (a) © Steve Gschmeissner/Science Source (b) © Biophoto Associates/Science Source

microtubules of the cilium. Centrosomes are also involved in the process of pulling duplicated chromosomes apart, as discussed in section 3.5.

Sperm cells are the only cells in the body that have **flagella.** The flagellum is a single, whiplike structure that propels the sperm through its environment. Like the motile cilia, a flagellum is composed of microtubules with a "9 + 2" arrangement. The subject of sperm motility by means of flagella is considered with the reproductive system in chapter 20.

Microvilli

In areas of the body that are specialized for rapid diffusion, the surface area of the plasma membranes may be increased by numerous folds called **microvilli.** The rapid passage of the products of digestion across the epithelial membranes in the intestine, for example, is aided by these structural adaptations. In the intestine, the surface area of the apical membranes (the part facing the lumen) is increased by the numerous tiny fingerlike projections (fig. 3.6). Similar microvilli are found in the epithelia of the kidney tubules, which must reabsorb various molecules that are filtered out of the blood.

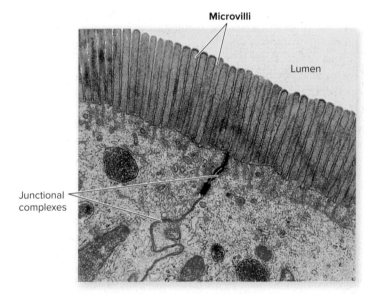

Microvilli

Lumen

Junctional complexes

FIGURE 3.6   **Microvilli in the small intestine.**   Microvilli are seen in this colorized electron micrograph, which shows two adjacent cells joined together by junctional complexes (chapter 6; see fig. 6.22).   AP|R   © Dennis Kunkel/Phototake

CHECKPOINTS

1a. Describe the structure of the plasma membrane.

1b. Describe the structure and function of cilia, flagella, and microvilli.

2a. Describe the different ways that cells can engulf materials in the extracellular fluid.

2b. Explain the process of exocytosis.

3.2 CYTOPLASM AND ITS ORGANELLES

Many of the functions of a cell are performed by structures called organelles. Among these are the lysosomes, which contain digestive enzymes, and the mitochondria, where most of the cellular energy is produced. Other organelles participate in the synthesis and secretion of cellular products.

LEARNING OUTCOMES

After studying this section, you should be able to:

3. Describe the structure and functions of the cytoskeleton, lysosomes, peroxisomes, mitochondria, and ribosomes.

4. Describe the structure and functions of the endoplasmic reticulum and Golgi complex, and explain how they interact.

Cytoplasm and Cytoskeleton

The material within a cell (exclusive of that within the nucleus) is known as **cytoplasm.** Cytoplasm contains structures called **organelles** that are visible under the microscope, and the fluid-like **cytosol** that surrounds the organelles. When viewed in a microscope without special techniques, the cytoplasm appears to be uniform and unstructured. However, the cytosol is not a homogeneous solution; it is, rather, a highly organized structure in which protein fibers—in the form of *microtubules* and *microfilaments*—are arranged in a complex latticework surrounding the membrane-bound organelles. Using fluorescence microscopy, these structures can be visualized with the aid of antibodies against their protein components (fig. 3.7). The interconnected microfilaments and microtubules are believed to provide structural organization for cytoplasmic enzymes and support for various organelles.

The latticework of microfilaments and microtubules is said to function as a **cytoskeleton** (fig. 3.8). The structure of this "skeleton" is not rigid; it is capable of quite rapid movement and reorganization. Contractile proteins—including actin and myosin, which are responsible for muscle contraction—are associated with the microfilaments and

microtubules in most cells. These structures aid in amoeboid movement, for example, so that the cytoskeleton is also the cell's "musculature." *Microtubules,* which are polymers of tubulin proteins, form a "track" along which motor proteins move their cargo through the cytoplasm. Additionally, microtubules form the mitotic and meiotic *spindles* that pull chromosomes away from each other in cell division. Microtubules also form the central parts of cilia and flagella and contribute to the structure and movements of these projections from the cells.

The cytoskeleton forms an amazingly complex "railway" system in a cell, on which large organelles (such as the nucleus), smaller membranous organelles (such as vesicles), and large molecules (including certain proteins and messenger RNA) travel to different and specific destinations. The molecular motors that move this cargo along their cytoskeletal tracks are the proteins *myosin* (along filaments of actin) and *kinesins* and *dyneins* (along microtubules). One end of these molecular motors attaches to their cargo while the other end moves along the microfilament or microtubule. For example, vesicles are moved in an axon (nerve fiber) toward its terminal by kinesin,

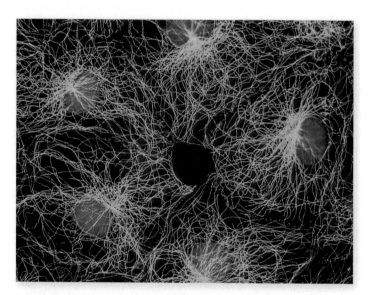

FIGURE 3.7 Immunofluorescence micrograph showing microtubules. In these fibroblast cells, the microtubules are green and the nuclei are blue. **AP|R** © Dr. Jan Schmoranzer/Science Source

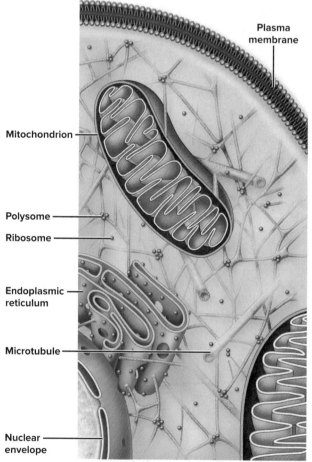

FIGURE 3.8 The formation of the cytoskeleton by microtubules. Microtubules are also important in the motility (movement) of the cell and movement of materials within the cell. **AP|R**

(Figure 3.8 labels: Plasma membrane, Mitochondrion, Polysome, Ribosome, Endoplasmic reticulum, Microtubule, Nuclear envelope)

while other vesicles can be transported in the opposite direction along the microtubule by dynein.

The cytoplasm of some cells contains stored chemicals in aggregates called **inclusions.** Examples are *glycogen granules* in the liver, striated muscles, and some other tissues; *melanin granules* in the melanocytes of the skin; and *triglycerides* within adipose cells.

Lysosomes

After a phagocytic cell has engulfed the proteins, polysaccharides, and lipids present in a particle of "food" (such as a bacterium), these molecules are still kept isolated from the cytoplasm by the membranes surrounding the food vacuole. The large molecules of proteins, polysaccharides, and lipids must first be digested into their smaller subunits (including amino acids, monosaccharides, and fatty acids) before they can cross the vacuole membrane and enter the cytoplasm.

The digestive enzymes of a cell are isolated from the cytoplasm and concentrated within membrane-bound organelles called **lysosomes,** which contain more than 60 different enzymes. A *primary lysosome* is one that contains only digestive enzymes within an environment that is more acidic than the surrounding cytoplasm.

A primary lysosome may fuse with a food vacuole to form a *secondary lysosome* that now contains the engulfed extracellular material. The digestion of structures and molecules within a vacuole by the enzymes within lysosomes is a process known as **autophagy.** Extracellular material digested by this process includes potentially disease-causing bacteria. Unique double-membrane vacuoles can also form intracellularly to contain structures within the cell, including viruses. A vacuole of this kind, called an *autophagosome,* can fuse with a lysosome so that the lysosomal enzymes degrade the viruses. In these and other ways, autophagy contributes to immunity.

In addition, autophagosomes can engulf parts of the cytoplasm and various organelles before fusing with lysosomes. This type of autophagy is stimulated in a nonspecific way when a cell is deprived of nutrients, so that digestion within lysosomes can release molecules needed by the cell. However, autophagy also occurs when a cell is not deprived of nutrients. Under these conditions, autophagy selectively eliminates damaged organelles, such as peroxisomes and mitochondria (discussed shortly) that are toxic to the cell.

In a similar manner, potentially toxic aggregations of proteins within the cytoplasm may be contained within autophagosomes and digested by lysosomes. Additionally, digestion by lysosomal enzymes is needed for the proper turnover of glycogen and certain lipids; lack of a particular enzyme can result in the undue accumulation of these molecules in the cell (see the next Clinical Application box). The 2016 Nobel Prize in Physiology or Medicine was awarded to a Japanese scientist for his discoveries regarding the genes, mechanisms, and functions of autophagy.

Lysosomes have also been called "suicide bags" because a break in their membranes would release their digestive enzymes and thus destroy the cell. This happens normally in *programmed cell death* (or *apoptosis*), described in section 3.5. An example is the loss of tissues that must accompany embryonic development, when earlier structures are remodeled or replaced as the embryo matures.

CLINICAL APPLICATION

Autophagosomes engulf parts of the cell before fusing with lysosomes when a cell is deprived of nutrients, releasing molecules needed by the cell. Autophagy also occurs when a cell selectively eliminates damaged organelles such as peroxisomes and mitochondria that are toxic to the cell. Autophagy and digestion by lysosomal enzymes produces a normal turnover of glycogen and certain complex lipids stored in cells. When a genetic defect results in a missing or defective lysosomal enzyme, the molecule normally degraded by the enzyme accumulates within the lysosomes. This produces enlarged and dysfunctional lysosomes, and can cause tissue damage. Examples of the approximately 50 different *lysosomal storage diseases* include **Gaucher's disease, Tay-Sach's disease,** and **Pompe disease.**

CLINICAL INVESTIGATION   CLUES

George had a grandparent on each side that had Gaucher's disease, the most common lysosomal storage disease. The doctor ordered tests to see if George inherited the condition.

- What are lysosomes, and what are their functions?
- What causes lysosomal storage diseases?

Peroxisomes

Peroxisomes are membrane-bound organelles containing several specific enzymes that promote oxidative reactions. Although peroxisomes are present in most cells, they are particularly large and active in the liver.

Peroxisomes contain enzymes that remove hydrogen from particular organic molecules. Removal of hydrogen oxidizes the molecules, and the enzymes that promote these reactions are called *oxidases* (chapter 4, section 4.3). The hydrogen is transferred to molecular oxygen (O_2), forming *hydrogen peroxide* (H_2O_2). Peroxisomes are important in the metabolism of amino acids and lipids and the production of bile acids. Peroxisomes also oxidize toxic molecules, such as formaldehyde and alcohol. For example, much of the ethanol (alcohol) ingested in drinks is oxidized to acetaldehyde by liver peroxisomes.

The enzyme *catalase* within the peroxisomes prevents the excessive accumulation of hydrogen peroxide by catalyzing the reaction $2H_2O_2 \rightarrow 2H_2O + O_2$. Catalase is one of the fastest acting enzymes known (see chapter 4), and it is this reaction that produces the characteristic fizzing when hydrogen peroxide is poured on a wound.

Mitochondria

All cells in the body, with the exception of mature red blood cells, have from a hundred to a few thousand organelles called **mitochondria** (singular, **mitochondrion**). Mitochondria serve as sites for the production of most of the energy of cells (chapter 5, section 5.2).

Mitochondria vary in size and shape, but all have the same basic structure (fig. 3.9). Each mitochondrion is surrounded by an inner and external membrane, separated by a narrow intermembranous space. The external mitochondrial membrane is smooth, but the inner membrane is characterized by many folds, called *cristae,* which project like shelves into the central area (or *matrix*) of the mitochondrion. The cristae and the matrix compartmentalize the space within the mitochondrion and have different roles in the generation of cellular energy. The structure and functions of mitochondria will be described in more detail in the context of cellular metabolism in chapter 5.

Mitochondria can migrate through a cell, combine together in a process called *fusion,* and reproduce themselves in a process called *fission.* Indeed, mitochondria contain their own DNA. All of the mitochondria in a person's body are derived from those inherited from the mother's fertilized egg cell. Thus, all of a person's mitochondrial genes are inherited from the mother. Mitochondrial DNA is more primitive (consisting of a circular, relatively small, double-stranded molecule) than that found within the cell nucleus. For this and other reasons, many scientists believe that mitochondria evolved from separate organisms, related to bacteria, that invaded the ancestors of animal cells and remained in a state of symbiosis.

This symbiosis might not always benefit the host; for example, mitochondria produce superoxide radicals that can provoke an oxidative stress (chapters 5 and 19), and some scientists believe that accumulations of mutations in mitochondrial DNA may contribute to aging. There are more than 150 mutations of mitochondrial DNA presently known to contribute to different human diseases. Examples include *Leber's hereditary neuropathy,* where patients undergo sudden loss of vision as young adults, and *MELAS* (an acronym for myopathy, encephalopathy, lactic acidosis, and strokelike episodes), a disorder affecting many organ systems. Mitochondrial DNA has only 37 genes that code for only 13 proteins (as well as 2 rRNAs and 22 tRNAs). The proteins are needed for oxidative phosphorylation (chapter 5, section 5.2), an essential part of aerobic respiration performed by mitochondria. However, each mitochondrion contains approximately 1,500 proteins, most of which are coded by DNA in the cell nucleus. Because of this, mitochondrial diseases may be produced by mutations in nuclear as well as mitochondrial DNA.

Neurons obtain energy solely from aerobic cell respiration (a process that requires oxygen, described in chapter 5), which occurs in mitochondria. Thus, mitochondrial fission (division) and transport over long distances is particularly important in neurons, where axons can be up to 1 meter in length. Mitochondria can also fuse together, which may help to repair those damaged by "reactive oxygen species" generated within mitochondria (chapters 5 and 19).

Although mitochondria are needed for aerobic cell respiration and are thus essential for the life of the cell, the production of reactive oxygen species by mitochondria can kill the cell. Also, when mitochondria become damaged, they can harm their host cells through the leakage of toxic mitochondrial molecules into the cytoplasm. The cell protects itself by enclosing

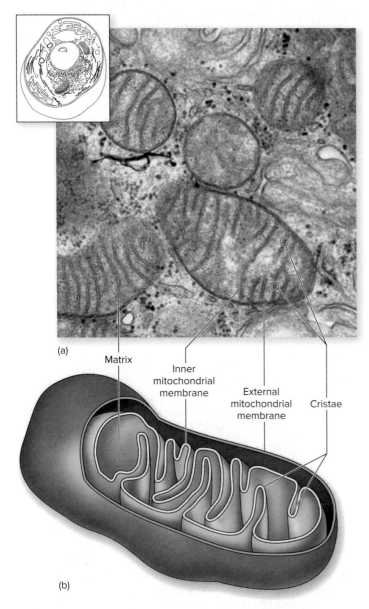

(a)

Matrix

Inner mitochondrial membrane

External mitochondrial membrane

Cristae

(b)

FIGURE 3.9 **The structure of mitochondria.** (*a*) Electron micrograph showing a few mitochondria. Notice the cristae formed from the inner mitochondrial membrane. (*b*) A diagram of a mitochondrion. **AP|R** (a) © EM Research Services, Newcastle University

the damaged mitochondria within an autophagosome, which can then fuse with a lysosome and be digested by the process of autophagy (discussed earlier).

Ribosomes

Ribosomes are often called the "protein factories" of the cell because it is here that proteins are produced according to the genetic information contained in messenger RNA (discussed in section 3.4). The ribosomes are quite tiny, about 25 nanometers in size, and can be found both free in the cytoplasm and located on the surface of an organelle called the endoplasmic reticulum.

Each ribosome consists of two subunits (fig. 3.10), which are designated 30S and 50S after their sedimentation rate in a centrifuge (this is measured in Svedberg units, from which the "S" is derived). Each of the subunits is composed of both ribosomal RNA and proteins. Contrary to earlier expectations of most scientists, the ribosomal RNA molecules serve as enzymes (called *ribozymes*) for many of the reactions in the ribosomes that are required for protein synthesis. Protein synthesis is covered in section 3.4, and the general subject of enzymes and catalysis is discussed in chapter 4.

Endoplasmic Reticulum

Most cells contain a system of membranes known as the **endoplasmic reticulum,** or **ER.** The ER may be either of two types: (1) a **rough endoplasmic reticulum** (fig. 3.11) or (2) a **smooth endoplasmic reticulum.** A rough endoplasmic reticulum bears ribosomes on its surface, whereas a smooth endoplasmic reticulum does not. The smooth endoplasmic reticulum serves a variety of purposes in different cells; it provides a site for enzyme reactions in steroid hormone production and inactivation, for example, and a site for the storage of Ca^{2+} in striated

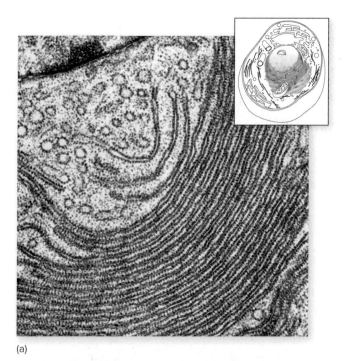

(a)

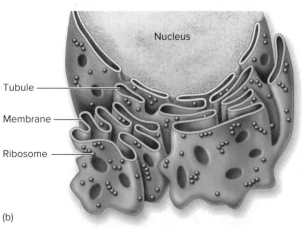

Nucleus

Tubule

Membrane

Ribosome

(b)

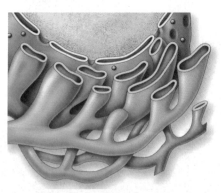

(c)

FIGURE 3.11   The endoplasmic reticulum. (*a*) An electron micrograph of a rough endoplasmic reticulum (about 100,000X). The rough endoplasmic reticulum (*b*) has ribosomes attached to its surface, whereas the smooth endoplasmic reticulum (*c*) lacks ribosomes. AP|R (a) © Oxford Scientific/Getty Images

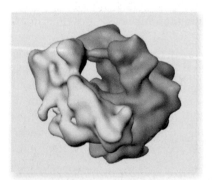

FIGURE 3.10   A ribosome is composed of two subunits. This is a model of the structure of a ribosome, showing the smaller (lighter) and larger (darker) subunits. The space between the two subunits accommodates a molecule of transfer RNA, needed to bring amino acids to the growing polypeptide chain. AP|R

muscle cells. A rough endoplasmic reticulum is abundant in cells that are active in protein synthesis and secretion, such as those of many exocrine and endocrine glands.

CLINICAL APPLICATION

Proliferation of the smooth endoplasmic reticulum of liver cells can occur in response to the abuse of alcohol and drugs. The smooth endoplasmic reticulum has numerous functions, including the enzymatic modification of toxic compounds into less toxic, more water-soluble derivatives that can be more easily excreted. Partly because of these enzymes, a person may gain **tolerance** to a drug, requiring a higher dose of the abused substance to produce the same effect. For example, a greater amount of the enzyme alcohol dehydrogenase will result in a faster elimination of alcohol. Tolerance to alcohol and barbiturates can overlap because of overlapping specificities of the detoxifying enzymes.

CLINICAL INVESTIGATION   CLUES

George's liver cells show an unusually extensive smooth endoplasmic reticulum.

- How does this relate to liver function, his symptoms, and what he told his doctor?

Golgi Complex

The **Golgi complex,** also called the **Golgi apparatus,** consists of a stack of several flattened sacs (fig. 3.12). This is something like a stack of pancakes, but the Golgi sac "pancakes" are hollow, with cavities called *cisternae* within each sac. One side of the stack faces the endoplasmic reticulum and serves as a site of entry of vesicles from the endoplasmic reticulum that contain cellular products. The other side of the stack faces the plasma membrane, and the cellular products somehow get transferred to that side. This may be because the products are passed from one sac to the next, probably in vesicles, until reaching the sac facing the plasma membrane. Alternatively, the sac that receives the products from the endoplasmic reticulum may move through the stack until reaching the other side.

By whichever mechanism the cell product is moved through the Golgi complex, it becomes chemically modified and then, in the sac facing the plasma membrane, is packaged into vesicles that bud off the sac. Depending on the nature of the cell product, the vesicles that leave the Golgi complex may become lysosomes, or secretory vesicles (in which the product is released from the cell by exocytosis), or may serve other functions. The scientists who discovered how secretory vesicles dock with their correct target membranes were awarded the 2013 Nobel Prize in Physiology or Medicine.

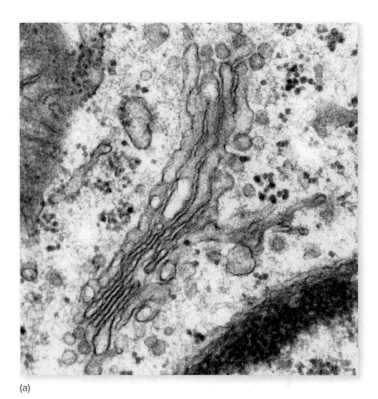

(a)

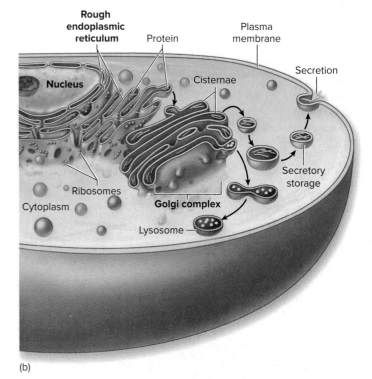

(b)

FIGURE 3.12   The Golgi complex. (*a*) An electron micrograph of a Golgi complex. Notice the formation of vesicles at the ends of some of the flattened sacs. (*b*) An illustration of the processing of proteins by the granular endoplasmic reticulum and Golgi complex. AP|R (a) © Oxford Scientific/Getty Images

The reverse of exocytosis is endocytosis, as previously described; the membranous vesicle formed by that process is an *endosome.* Some cellular proteins that were released by exocytosis are recycled by a pathway that is essentially the reverse of the one depicted in figure 3.12. This reverse pathway is called **retrograde transport,** because proteins within the extracellular fluid are brought into the cell and then taken to the Golgi apparatus and the endoplasmic reticulum. Also, some toxins, such as the cholera toxin, and proteins from viruses (including components of HIV) rely on retrograde transport for their ability to infect cells.

CHECKPOINTS

3a. Explain why microtubules and microfilaments can be thought of as the skeleton and musculature of a cell.

3b. Describe the functions of lysosomes and peroxisomes.

3c. Describe the structure and functions of mitochondria.

3d. Explain how mitochondria can provide a genetic inheritance derived only from the mother.

3e. Describe the structure and function of ribosomes.

4. Distinguish the two types of endoplasmic reticulum and explain the relationship between the endoplasmic reticulum and the Golgi complex.

3.3 CELL NUCLEUS AND GENE EXPRESSION

The nucleus is the organelle that contains the DNA of a cell. A gene is a length of DNA that codes for the production of a specific polypeptide chain. In order for genes to be expressed, they must first direct the production of complementary RNA molecules. That process is called genetic transcription.

LEARNING OUTCOMES

After studying this section, you should be able to:

5. Describe the structure of the nucleus and of chromatin, and distinguish between different types of RNA.

6. Explain how DNA directs the synthesis of RNA in genetic transcription.

Most cells in the body have a single **nucleus** (fig. 3.13). Exceptions include skeletal muscle cells, which have many nuclei, and mature red blood cells, which have none. The nucleus is enclosed by two membranes—an inner membrane and an outer membrane—that together are called the **nuclear envelope.** The outer membrane is continuous with the endoplasmic reticulum in the cytoplasm. At various points, the inner and outer membranes are fused together by structures called *nuclear pore complexes.* These structures function as rivets, holding the two

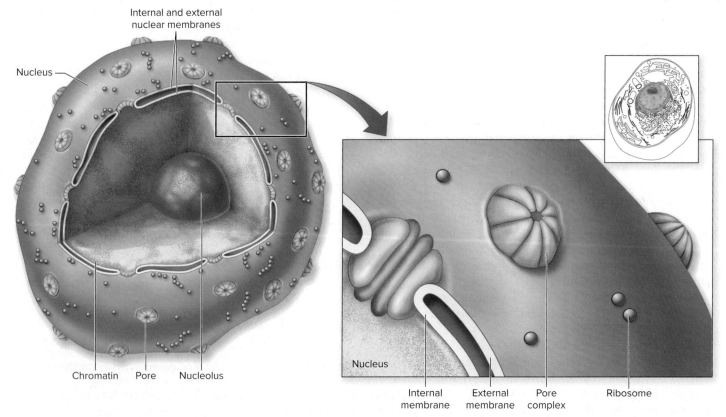

FIGURE 3.13 **The nuclear membranes and pores.** A diagram showing the inner and outer nuclear membranes and the nuclear pore complexes. The nucleolus within the nucleus is also shown. **AP|R**

membranes together. Each nuclear pore complex has a central opening, the *nuclear pore* (fig. 3.13), surrounded by interconnected rings and columns of proteins. Small molecules may pass through the complexes by diffusion, but movement of protein and RNA through the nuclear pores is a selective process that requires energy for proteins to ferry their cargo into and out of the nucleus.

Transport of specific proteins from the cytoplasm into the nucleus through the nuclear pores may serve a variety of functions, including regulation of gene expression by hormones (see chapter 11). Transport of RNA out of the nucleus, where it is formed, is required for gene expression. As described in this section, genes are regions of the DNA within the nucleus. Each gene contains the code for the production of a particular type of RNA called messenger RNA (mRNA). As an mRNA molecule is transported through the nuclear pore, it becomes associated with ribosomes that are either free in the cytoplasm or associated with the rough endoplasmic reticulum. The mRNA then provides the code for the production of a specific type of protein.

The primary structure of the protein (its amino acid sequence) is determined by the sequence of bases in mRNA. The base sequence of mRNA has been previously determined by the sequence of bases in the region of the DNA (the gene) that codes for the mRNA. **Genetic expression** therefore occurs in two stages: first **genetic transcription** (synthesis of RNA) and then **genetic translation** (synthesis of protein).

Each nucleus contains one or more dark areas (fig. 3.13). These regions, which are not surrounded by membranes, are called **nucleoli.** The DNA within the nucleoli contains the genes that code for the production of ribosomal RNA (rRNA).

Genome and Proteome

The term **genome** can refer to all of the genes of a particular individual or all of the genes of a particular species. **Genes** have historically been defined as regions of DNA that code (through production of mRNA) for polypeptide chains. The *Human Genome Project* was completed in 2001, and currently scientists have found about 20,000 protein-coding genes in humans, far fewer than the number previously assumed. Scientists were surprised to learn that the protein-coding regions comprise little more than 1% of the genome. The rest was initially thought to be "junk DNA," but this idea was erroneous. A project called *ENCODE* (for the Encyclopedia of DNA Elements) is cataloguing the function of these elements. This project has shown that, although only about 1% of the genome codes for proteins, 62% is transcribed into some form of RNA and another 20% performs other functions. A more inclusive definition of "gene" would include DNA regions that encode any type of RNA.

Part of the genome codes for RNA molecules such as microRNA (discussed shortly), which attenuate the translation of mRNA and thereby regulate gene expression. Part serves as "promoter" regions, which are sites that participate in the regulation of nearby genes. Other stretches of DNA serve as "enhancer" regions that regulate the expression of distant genes. Some other stretches of DNA constitute *jumping genes,* or *retrotransposons,* that make copies of themselves during cell division and insert these copies randomly into the genome, thereby changing genetic expression. In the performance of these and other functions, most of the genome that lies outside of the historical definition of genes (the protein-coding regions) is transcribed into RNA at some time.

Until recently, it was believed that one gene could be defined as coding for one polypeptide chain (recall that some proteins consist of two or more polypeptide chains; see fig. 2.28*e,* for example). However, each cell produces well over 100,000 different proteins, so the number of proteins greatly exceeds the number of genes. The term **proteome** has been coined to refer to all of the proteins produced by the genome. This concept is complicated because, in a given cell, some portion of the genome is inactive. There are proteins produced by a neuron that are not produced by a liver cell, and vice versa. Further, a given cell will produce different proteins at different times, as a result of signaling by hormones and other regulators.

So, how does a gene produce more than one protein? This is not yet completely understood. Part of the answer may include the following: (1) a given RNA coded by a gene may be cut and spliced together in different ways (see fig. 3.17), in a process called *alternative splicing,* discussed shortly; (2) a particular polypeptide chain may associate with different polypeptide chains to produce different proteins; (3) many proteins have carbohydrates or lipids bound to them, which alter their functions. There is also a variety of *posttranslational modifications* of proteins (made after the proteins have been formed), including chemical changes such as methylation and phosphorylation, as well as the cleavage of larger polypeptide chain parent molecules into smaller polypeptides with different actions. Scientists have estimated that an average protein has at least two or three of such posttranslational modifications. These variations of the polypeptide products of a gene allow the human proteome to be many times larger than the genome.

Part of the challenge of understanding the proteome is identifying all of the proteins. This is a huge undertaking, involving many laboratories and biotechnology companies. The function of a protein, however, depends not only on its composition but also on its three-dimensional, or tertiary, structure (see fig. 2.28*d*) and on how it interacts with other proteins. The study of genomics, proteomics, and related disciplines will challenge scientists into the foreseeable future and, it is hoped, will yield important medical applications in the coming years.

Chromatin

DNA is composed of four different nucleotide subunits that contain the nitrogenous bases adenine, guanine, cytosine, and thymine. These nucleotides form two polynucleotide chains, joined by complementary base pairing and twisted to form a double helix. This structure is discussed in chapter 2 and illustrated in figures 2.32 and 2.33.

The DNA within the cell nucleus is combined with protein to form **chromatin,** the threadlike material that makes up the chromosomes. Much of the protein content of chromatin is of a

CLINICAL APPLICATION

It is estimated that only about 300 genes out of a total of about 20,000 are active in any given cell. This is because each cell becomes specialized in a process called *differentiation.* The differentiated cells of an adult are derived, or "stem from," those of the embryo. **Embryonic stem cells** can become any cell in the body—they are said to be *pluripotent.* The chromatin in embryonic stem cells is mostly euchromatin, with an open structure that permits its genes to be expressed. As development proceeds, more condensed regions of heterochromatin appear as genes become silenced during differentiation. **Adult stem cells** can differentiate into a range of specific cell types, but they are not normally pluripotent. For example, the bone marrow of an adult contains two types of adult stem cells. These include **hematopoietic stem cells,** which can form the blood cells, and **mesenchymal stem cells,** which can differentiate into osteocytes (bone cells), chondrocytes (cartilage cells), and adipocytes (fat cells). **Neural stem cells** have been identified in the adult nervous system. These can migrate to particular locations and differentiate into specific neuron and neuroglial cell types in these locations. Adult stem cells are also found in other organs, including the epithelium of the skin and intestine. The heart even seems to contain adult stem cells, although they are insufficient to repair the damage of myocardial infarction (heart attack).

type known as *histones.* Histone proteins are positively charged and organized to form spools, about which the negatively charged strands of DNA are wound. Each spool consists of two turns of DNA, comprising 146 base pairs, wound around a core of histone proteins. This spooling creates particles known as **nucleosomes** (fig. 3.14).

Chromatin that is active in genetic transcription (RNA synthesis) is in a relatively extended form known as **euchromatin.** By contrast, **heterochromatin** is highly condensed and forms blotchy-looking areas in the nucleus. As long as an area of chromatin is in the condensed state, its genes are inactive.

In the euchromatin, genes may be activated or repressed at different times. This is believed to be accomplished by chemical changes in the histones. Such changes include acetylation (the addition of two-carbon-long chemical groups), which turns on genetic transcription, and deacetylation (the removal of those groups), which stops the gene from being transcribed. The acetylation of histone proteins produces a less condensed, more open configuration of the chromatin in specific locations (fig. 3.15), allowing the DNA to be "read" by transcription factors (those that promote RNA synthesis, described next). These and other histone modifications, together with methylation of DNA, regulate the ability of transcription factors to gain access to the DNA and promote gene expression.

RNA Synthesis

Each gene is a stretch of DNA that is several thousand nucleotide pairs long. The DNA in a human cell contains over 3 billion base pairs—enough to code for at least 3 million proteins. Because the average human cell contains fewer proteins than this (30,000 to 150,000 different proteins), it follows that only a fraction of the DNA in each cell is used to code for proteins.

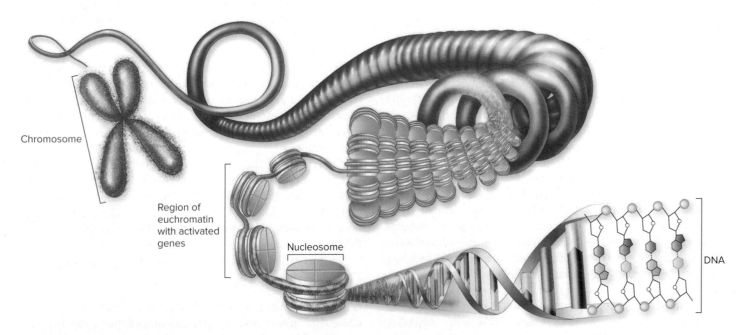

Chromosome

Region of euchromatin with activated genes

Nucleosome

DNA

FIGURE 3.14   The structure of chromatin. Part of the DNA is wound around complexes of histone proteins, forming particles known as nucleosomes. AP|R

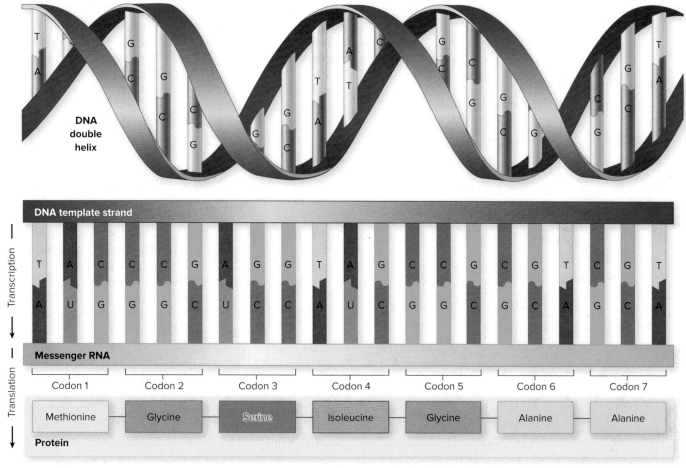

DNA double helix

DNA template strand

Transcription

| T | A | C | C | C | G | A | G | G | T | A | G | C | C | G | C | G | T | C | G | T |

| A | U | G | G | G | C | U | C | C | A | U | C | G | G | C | G | C | A | G | C | A |

Messenger RNA

Translation

| Codon 1 | Codon 2 | Codon 3 | Codon 4 | Codon 5 | Codon 6 | Codon 7 |

| Methionine | Glycine | Serine | Isoleucine | Glycine | Alanine | Alanine |

Protein

FIGURE 3.19 **Transcription and translation.** The genetic code is first transcribed into base triplets (codons) in mRNA and then translated into a specific sequence of amino acids in a polypeptide. AP|R

See the *Test Your Quantitative Ability* section of the **Review Activities** at the end of this chapter.

Each synthetase enzyme recognizes its amino acid and joins it to the tRNA that bears a specific anticodon. The cytoplasm of a cell thereby contains tRNA molecules that are each bonded to a specific amino acid, and each of these tRNA molecules is capable of bonding with a specific codon in mRNA via its anticodon base triplet.

Formation of a Polypeptide

The anticodons of tRNA bind to the codons of mRNA as the mRNA moves through the ribosome. Because each tRNA molecule carries a specific amino acid, the joining together of these amino acids by peptide bonds creates a polypeptide whose amino acid sequence has been determined by the sequence of codons in mRNA.

Two tRNA molecules containing anticodons specific to the first and second mRNA codons enter a ribosome, each carrying its own specific amino acid. After anticodon-codon binding between the tRNA and mRNA, the first amino acid detaches from its tRNA and bonds to the second amino acid, forming a dipeptide attached to the second tRNA. While this occurs, the mRNA moves down a distance of one codon within the

ribosome, allowing the first tRNA (now minus its amino acid) to detach from the mRNA. The second tRNA with its dipeptide is thereby moved down one position in the ribosome. A third tRNA, bearing its specific amino acid, then attaches by its anticodon to the third codon of the mRNA. The previously formed dipeptide is now moved to the amino acid carried by the third tRNA as the mRNA again moves a distance of one codon within the ribosome. This is followed by the release of the second tRNA (minus its dipeptide), as the third tRNA, which now carries a tripeptide, moves up a distance of a codon in the ribosome. A polypeptide chain, bound to one tRNA, thereby grows as new amino acids are added to its growing tip (fig. 3.21). This process continues until the ribosome reaches a "stop" codon in the mRNA, at which point genetic translation is terminated and the fully formed polypeptide is released from the last tRNA.

As the polypeptide chain grows in length, interactions between its amino acids cause the chain to twist into a helix (secondary structure) and to fold and bend upon itself (tertiary structure). At the end of this process, the new protein detaches from the tRNA as the last amino acid is added. Although, under ideal conditions, the newly formed polypeptide chain could fold correctly to produce its proper tertiary structure, this might not happen in

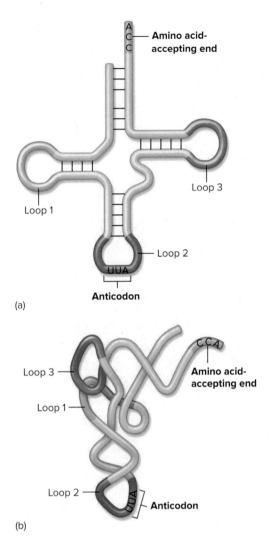

(a)

(b)

FIGURE 3.20   **The structure of transfer rna (tRna).** (*a*) A simplified cloverleaf representation and (*b*) the three-dimensional structure of tRNA.

the cell. For example, one region of the newly forming polypeptide chain might improperly interact with another region before the chain has fully formed. Also, similar proteins in the vicinity might aggregate with the newly formed polypeptide to produce toxic complexes. Such inappropriate interactions are normally prevented by **chaperones,** which are proteins that help the polypeptide chain fold into its correct tertiary structure as it emerges from the ribosome. Chaperone proteins are also needed to help different polypeptide chains come together in the proper way to form the quaternary structure of particular proteins (chapter 2).

Many proteins are further modified after they are formed. These modifications occur in the rough endoplasmic reticulum and Golgi complex.

Functions of the Endoplasmic Reticulum and Golgi Complex

Proteins that are to be used within the cell are likely to be produced by polyribosomes that float freely in the cytoplasm,

unattached to other organelles. If the protein is to be secreted by the cell, however, it is made by mRNA-ribosome complexes that are located on the rough endoplasmic reticulum. The membranes of this system enclose fluid-filled spaces called *cisternae* (sing. *cisterna*) into which the newly formed proteins may enter. Once in the cisternae, the structure of these proteins is modified in specific ways.

When proteins destined for secretion are produced, the first 30 or so amino acids are primarily hydrophobic. This *leader sequence* is attracted to the lipid component of the membranes of the endoplasmic reticulum. As the polypeptide chain elongates, it is "injected" into the cisterna within the endoplasmic reticulum. The leader sequence is, in a sense, an "address" that directs secretory proteins into the endoplasmic reticulum. Once the proteins are in the cisterna, the leader sequence is enzymatically removed so that the protein cannot reenter the cytoplasm (fig. 3.22).

The processing of the hormone insulin can serve as an example of the changes that occur within the endoplasmic reticulum. The original molecule enters the cisterna as a single polypeptide composed of 109 amino acids. This molecule is called *preproinsulin*. The first 23 amino acids serve as a leader sequence that allows the molecule to be injected into the cisterna within the endoplasmic reticulum. The leader sequence is then quickly removed, producing a molecule called *proinsulin*. The remaining chain folds within the cisterna so that the first and last amino acids in the polypeptide are brought close together. Enzymatic removal of the central region produces two chains—one of them 21 amino acids long, the other 30 amino acids long—that are subsequently joined together by disulfide bonds (fig. 3.23). This is the form of insulin that is normally secreted from the cell.

Secretory proteins do not remain trapped within the rough endoplasmic reticulum. Instead, they are transported to another organelle within the cell—the Golgi complex (Golgi apparatus), as previously described. This organelle serves three interrelated functions:

1. Proteins are further modified (including the addition of carbohydrates to some proteins to form *glycoproteins*) in the Golgi complex.
2. Different types of proteins are separated according to their function and destination in the Golgi complex.
3. The final products are packaged and shipped in vesicles from the Golgi complex to their destinations (see fig. 3.12).

In the Golgi complex, for example, proteins that are to be secreted are separated from those that will be incorporated into the plasma membrane and from those that will be introduced into lysosomes. Each is individually packaged in a membrane-enclosed vesicle and sent to its proper destination.

Protein Degradation

Proteins within a cell have numerous regulatory functions. Many proteins are enzymes, which increase the rate of specific chemical reactions (chapter 4). This can have diverse